Common and Scientific Names of Aquatic Invertebrates from the United States and Canada: Cnidaria and Ctenophora

Publication of this book was made possible by grants from the

Department of Commerce / NOAA / National Ocean Service
National Centers for Coastal Ocean Science
Marine Sanctuaries Division / Office of Response and Restoration

Department of Commerce / NOAA / National Marine Fisheries Service

Department of Commerce / NOAA / Office of Oceanic and Atmospheric Research

Department of Commerce / NOAA / National Environmental Satellite, Data, and Information Service

Department of the Interior / U.S. Geological Survey
Biological Resources Discipline

Department of the Interior / National Park Service

Travel and institutional support was extended to the Committee by the

American Fisheries Society

and the

Authors' Institutions

American Fisheries Society Special Publication 28

Common and Scientific Names of Aquatic Invertebrates from the United States and Canada:

Cnidaria and Ctenophora

Second Edition, 2002

Stephen D. Cairns, *Chair*
Dale R. Calder, Anita Brinckmann-Voss, Clovis B. Castro,
Daphne G. Fautin, Philip R. Pugh, Claudia E. Mills, Walter C. Jaap,
Mary N. Arai, Stephen H. D. Haddock, and Dennis M. Opresko

Committee on Common and Scientific Names of Cnidaria and Ctenophora

American Fisheries Society
Bethesda, Maryland
2002

The American Fisheries Society Special Publication series is a registered serial. A suggested citation format follows.

Cairns, S. D., D. R. Calder, A. Brinckmann-Voss, C. B. Castro, D. G. Fautin, P. R. Pugh, C. E. Mills, W. C. Jaap, M. N. Arai, S. H. D. Haddock, and D. M. Opresko. 2002. Common and scientific names of aquatic invertebrates from the United States and Canada: Cnidaria and Ctenophora. 2nd edition. American Fisheries Society, Special Publication 28, Bethesda, Maryland.

Library of Congress Control Number 2002107362
ISBN 1-888569-39-5
ISSN 0097-0638

Printed in the United States of America on acid-free paper

American Fisheries Society
5410 Grosvenor Lane, Suite 110
Bethesda, Maryland 20814-2199
USA

CONTENTS

Part I

Part II

Part III

FOREWORD

The American Fisheries Society (AFS) Committee on Names of Aquatic Invertebrates (CNAI) was established by AFS President John J. Magnuson on 30 September 1981. The main goal of this committee is to achieve uniformity and avoid confusion in vernacular nomenclature of aquatic invertebrates. The present charge by the society to this committee is as follows:

> The committee shall be responsible for studying and reporting on matters concerning common and scientific names of aquatic invertebrates and shall prepare checklists of names to achieve uniformity and avoid confusion in nomenclature. The chairman shall be custodian of the master checklists. The committee shall coordinate its activities with those of other societies and organizations throughout the world. The committee shall be composed of members who are outstanding specialists in invertebrate taxonomy and nomenclature.

The CNAI has benefited substantially from the long experience and decisions reached by the AFS Names of Fishes Committee. The Names of Fishes Committee was originally appointed in 1933 as the result of a resolution adopted by the society to form a permanent committee of experts in the field of ichthyology "to prepare and submit for publication a list of common names of fishes corresponding to the accepted scientific names." Because the AFS membership does not include much of the invertebrate taxonomic expertise needed to develop a comprehensive list of common and scientific names of aquatic invertebrates from the United States and Canada, it was decided to enlist the cooperation of other professional societies and taxonomic specialists to accomplish this goal.

The American Fisheries Society gratefully acknowledges one more significant contribution by Dr. Stephen D. Cairns, a member of CNAI and chair of the Committee on Common Names of Cnidaria and Ctenophora, to the series of volumes on Common and Scientific Names of Aquatic Invertebrates from the United States and Canada. Eleven years after publication of his initial volume, Dr. Cairns is to be commended for bringing together his committee of 10 specialists and completing the second edition of the *Cnidaria and Ctenophora*. Their collective efforts are appreciated.

Committee on Names of Aquatic Invertebrates

Donna D. Turgeon, *Chair*

Cynthia Ahearn
Edward L. Bousfield
Stephen Cairns
David Camp
Kathryn Coates
Linda Cole
Kristian Fauchald
Brian Kensley
Pat Mclaughlin
Shirley Pomponi
Andrew Robertson
Lynne Starnes
Fred G. Thompson
Mark Wetzel

ALPHABETICAL LIST OF FAMILIES

INTRODUCTION

The purpose of this volume is to provide a checklist of species and to recommend selected common names for North American Cnidaria and Ctenophora, thereby achieving uniformity and avoiding confusion in vernacular nomenclature. It is not intended as a taxonomic revision or an imposition of scientific names. Common names have the potential to be stabilized by general agreement; scientific names, on the other hand, often change with advancing knowledge. Unlike the list of North American fishes (Robins et al. 1991), it has not been possible to provide common names for all species; in fact, only 21% of the 1,369 taxa listed here have been given vernacular names. Whereas many of the stony corals and anemones do have common names, most cnidarians are too small, inconspicuous, uncommon, or poorly known to be so recognized. Should these rare species become objects of greater interest, common names may be applied in subsequent revisions of this list.

In addition to stabilizing common name nomenclature, we hope that this list will heighten public awareness of the diversity and wide distribution of cnidarians in North America, help identify taxonomic groups in need of systematic revision, and serve as a preliminary guide to the literature required for the identification of species.

This second edition of the cnidarian and ctenophoran list represents a net increase of only 57 species in the intervening decade, but one must also consider that several dozen species listed in the first edition have been removed because of synonymy or having been found to be extralimital.

Area of Coverage

The present list purports to include all species of Cnidaria known from the fresh waters of North America north of Mexico and species of marine Cnidaria and Ctenophora that inhabit contiguous shore waters on or above the continental shelf, to a depth of 200 m. Included in this geographic area are the northern Gulf of Mexico to the Rio Grande and the Arctic shore waters north of Alaska and Canada. The coverage for primarily pelagic taxa, such as Narcomedusae, Trachymedusae, Siphonophora, and Ctenophora, the occurrences of which seldom directly depend on bottom depth, is extended to 320 km from the coast but does not include the Bahamas or Cuba. Otherwise, records of the Anthozoa and those hydromedusae linked to benthic polyp stages are limited to depths shallower than 200 m. Species included are those whose occurrence in this region has been authenticated in published accounts or verified by a specialist, based on specimens in an established research collection.

The list provides a general guide to distribution designated by the following symbols: A (Atlantic Ocean, including the Gulf of Mexico), P (Pacific Ocean), Ac (Arctic Ocean), F (fresh water), I (introduced), and E (estuarine). Grainger's (1971) definition of the Arctic Ocean is followed; to the west, it includes the waters north of the Bering Strait, and to the east, it includes all waters west of the eastern end of Hudson Strait (including Hudson Bay, Ungava Bay, Frobisher Bay, and Cumberland Sound) and the waters north of the Arctic Circle in Davis Strait. An estuary is defined as "a semi-enclosed coastal body of water having a free connection with the open sea and within which the sea water is measurably diluted with fresh water runoff" (Pritchard 1955: 1).

Of the 1,369 taxa reported from North America, 66.5% occur in the Atlantic, 50.5% in the Pacific, and 7.5% in the Arctic Ocean. Sixteen species, primarily the Hydridae and Limnomedusae, are found in freshwater habitats, seven species are endemic to estuaries, and eleven are considered to be introduced. (The total is more than 100% because 300-plus species occur in more than one ocean.) These percentages are virtually the same as those reported in the first edition, except for a slight increase in the number of species found off the Pacific coast of North America.

Principles Governing Selection of Common Names

1. *A primary vernacular name shall be accepted for each species or taxonomic unit included.* Alternative published names may be listed in order of prominence. Rationale for selection of the primary name and etymologies may be indicated.

2. *No two species on the list shall have the same primary vernacular name.* Commonly used names of extralimital species should be avoided wherever possible.

3. *The expression "common" as part of an invertebrate's name shall be avoided if possible.* Use of adjectives that also describe age or size (e.g., little, small, big, fat) and thus may have dual meanings shall be avoided as part of an invertebrate's name wherever possible.

4. *Simplicity in names is favored.* Hyphens and suffixes shall be omitted (e.g., ostrichplume hydroid) except where they are orthographically essential (e.g., by-the-wind sailor), have a special meaning, or are necessary to avoid possible misunderstanding. Compounded modifying words, including those that denote paired structures, usually should be treated as singular nouns in apposition with a group name (e.g., horsetail hydroid, bottlebrush hydroid), but a plural modifier should usually be placed in adjectival form (e.g., mottled anemone, branched finger coral) unless its plural nature is obvious (e.g., reticulate anemone). Preference shall be given to names that are short and euphonious.

The compounding of brief, familiar words into a single name, written without a hyphen, may in some cases promote clarity and simplicity (e.g., snailfur, jellyfish), but the habitual practice of combining words, especially those that are lengthy, awkward, or unfamiliar, is to be avoided. A guide for spelling compound names is set forth in the fifth edition of the *CBE Style Manual* (CBE Style Manual Committee 1983).

5. *Common names shall not be capitalized in text use except for those elements that are proper names* (e.g., lettuce coral, but San Francisco anemone).

6. *Names intended to honor persons* (e.g., Williams anemone) *are discouraged in that they are without descriptive value.* In some large groups, identical specific patronymics in scientific names (sometimes honoring different persons) exist in related genera, and use of a patronymic in the common name is confusing. Some patronymics are already well established in the literature, agency regulations, and industry, however, and thus are best retained (e.g., Lights anemone). Apostrophes should be deleted.

7. *Only clearly defined and well-marked taxonomic entities (usually species) shall be assigned common names.* Most subspecies are not suitable subjects for common names, but those forms that are so different in appearance (not just in geographic distribution) as to be distinguished readily by laypeople or for which a common name constitutes a significant aid in communication may merit separate names. There is a wide divergence of opinion concerning the criteria for recognition of subspecies; therefore, they are usually not given common names. Subspecies have importance in evolutionary study but are rarely of significance to laypeople or in those aspects of biological endeavor in which common names are of concern. Exceptions to these general cases include invertebrates such as the freshwater mollusk yellow blossom *Epioblasma florentina florentina*, which is listed by the U.S. Department of the Interior as endangered. The common name for the species should apply to all subspecies of a taxon and may be appropriately modified by those treating subspecies. The practice of adding geographic modifiers to designate regional populations makes for a cumbersome terminology.

Hybrids in general are not named. The established common name of a hybrid, if important, is indicated by a footnote. Cultured varieties, phases, and morphological variants also are not named even though they are important in commercial trade of aquarium animals.

8. *The common name shall not be intimately tied to the scientific name.* Thus, the progress of scientific nomenclature, which is governed by the *International Code of Zoological Nomenclature* (International Commission on Zoological Nomenclature 1999), does not entail changing common names. The practice of applying a name to each genus, a modifying name for each species, and still another modifier for each subspecies, while appealing in its simplicity, has the defect of inflexibility. If an invertebrate is transferred from genus to genus, or shifted from species to subspecies or vice versa, the common name should remain unaffected. It is not a primary function of common names to indicate relationship. When two or more taxonomic groups (e.g., nominal species) are found to be identical, one name shall be adopted for the combined group.

This principle is regarded not only as fundamental to the achievement of stability, but also as essential to the development of a true vernacular nomenclature.

9. *Names shall not violate the tenets of good taste.*

The foregoing principles are largely procedural precepts. Those given below are criteria that may aid in the selection of suitable names.

10. *Colorful, romantic, fanciful, metaphorical, and otherwise distinctive and original names are especially appropriate.* Examples include Portuguese man o' war, lion's mane, cowardly anemone, sea pansy, fire coral, and golfball coral.

11. *American Indian or other truly vernacular names are welcome for adoption as common names.* Native names in current use in other AFS lists include the mollusk taxa cayuse physa, geoduck, and quahog. In addition to aboriginal names, names of American invertebrates have been derived from people of non-English-speaking extraction (e.g., Spanish: abalone, aglaja).

12. *Commonly employed names adapted from traditional English usage (e.g., coral, anemone, jellyfish) are given considerable latitude in taxonomic placement.* Adherence to customary English practice is to be preferred if this does not conflict with the broad general use of another name. Many English names, however,

have been applied to similar-appearing but often distantly related invertebrates in North America. We find "coral" in use for representatives of several orders, including the Antipatharia, Alcyonacea, and Scleractinia. "Jellyfish" is applied to members of the classes Cubozoa and Scyphozoa and the hydrozoan orders Anthomedusae, Limnomedusae, Leptomedusae, Narcomedusae, and Trachymedusae. For widely known species, the Committee believes it preferable to recognize and adopt general use than to adopt bookish or pedantic substitutes. Thus, established practice should outweigh consistency with original English usage.

13. *Structural attributes, color, and color pattern are desirable and are in common use in forming names.* Annulate, corky, hairy, leathery, striate, prickly, velvety, whitecross, and a multitude of other descriptors decorate invertebrate names. Efforts should be made to select terms that are descriptively accurate and to minimize repetition of those most frequently employed (e.g., white, black, spotted, banded).

Following tradition in North American invertebrate zoology, we have attempted to restrict use of line or stripe to longitudinal marks that parallel the body axis, reserving bar or band to vertical or transverse markings.

14. *Ecological characteristics are useful in making good names.* They, too, should be properly descriptive. Terms such as reef, pond, coral, sand, rock, and freshwater are well known in invertebrate names.

15. *Geographic distribution provides suitable adjectival modifiers.* Poorly descriptive or misleading geographic characterizations should be corrected unless they are too deeply entrenched in current usage. In the interest of brevity, it is usually possible to delete words such as lake, river, or ocean in the names of species (e.g., Caribbean horn coral, not Caribbean Sea horn coral).

16. *Generic names may be employed outright* (e.g., tubularia, hydra) *or in modified form as common names.* Once adopted, such names should be maintained even if the generic name is changed. These vernaculars should be written in Roman and without capitalization. Brevity and euphony are of special importance for names of this type.

17. *The duplication of common names of invertebrates and other organisms should be avoided if possible,* but names in wide general use need not be rejected on this basis alone. The name pansy is commonly applied to the herbaceous plant *Viola* and also to certain sea pens. Similarly, olive is employed both for the fruit of a tree, *Olea europea,* and for various gastropods. On the basis of prevailing use, these names are admissible as invertebrate names.

Plan of List

The phylum name Coelenterata ("coelenterates") is sometimes used to refer to the phylum Cnidaria or the combined phyla Cnidaria and Ctenophora. We have, however, adopted the currently accepted and popular practice of placing Cnidaria (=Coelenterata) and Ctenophora in separate phyla; Cnidaria is preferred because Coelenterata, although named first, was originally proposed for a polyphyletic assemblage that included sponges.

The list of cnidarians and ctenophores is presented as a natural or phyletic sequence of classes, orders, and families, with the genera and species within each family arranged alphabetically. The higher classification of Cnidaria is complex, controversial, and frequently changing; the classification adopted here reflects the current taxonomic opinions of the specialists who compiled their respective parts of the list.

Authors and dates of establishment for scientific names are included because these are commonly needed by persons who may not have ready access to the original references. Use of the authority's name(s) reflects current interpretation of the *International Code of Zoological Nomenclature* (International Commission on Zoological Nomenclature 1999). In line with that code, the author's name follows the specific name directly (and without punctuation) if the species, when originally described, was assigned to the same genus in which it appears here; if the species was originally described in another genus, the author's name appears in parentheses. For example, Forsskål originally named the club hydroid *Hydra multicornis,* but it appears in the list as *Clava multicornis* (Forsskål, 1775).

In the checklist that follows, all new entries that depart in any way (scientific name, authority, occurrence, or common name) from the first edition are preceded by an asterisk (*); such entries, keyed to the page in the main list, are explained in Part II. A plus sign (+) before an entry indicates that although the entry is unchanged from the first edition, a comment will be found in Part II under that name.

Several orders of hydrozoans have complex life cycles, sometimes referred to as "alternation of generations" or metagenesis, consisting of alternating periods as an attached asexual polyp (hydroid), then as a free sexual medusa (or variations on this general theme). Many species have been independently described based on the polyp or medusa stage, and different species and generic names have often been assigned to the different stages of the same species. As a result, a dual nomenclature has developed in which, for example, the Anthoathecatae hydroid is the polyp equivalent of the Anthoathecatae hydromedusae. The taxonomy is slowly being resolved through patient rearing of species through complete life cycles. More names of species will undoubtedly be synonymized as life histories are recorded and each polyp is linked to its corresponding medusoid stage. Because the hydroid and medusoid stages of the same species occur in different ecological habitats and are usually studied by different specialists, it is not the intention of this list to attempt a taxonomic reconciliation of these names. Instead, the hydroid stage names within two orders (Anthoathecatae and Leptothecatae) are listed first, followed by a listing of the medusoid stage names pertaining to the coordinate hydromedusae. Those 55 species known from both hydroid and medusoid stages are therefore listed twice, and each name is preceded by a caret (^) to denote its additional listing. In some of the species with a known life cycle, only medusa or only hydroid may have been reported from U.S. or Canadian waters. For instance, the medusa of *Corymorpha forbesi* is listed because it is known from the Atlantic coast of the United States, but its hydroid is not listed because thus far it has been collected only from the Mediterranean. However, hydroids and hydromedusae raised in North American laboratories (but not yet collected in the field) are listed (e.g., *Zanclea bomala* is listed as a hydroid because it was collected from the Pacific coast of the United States, but it is also listed as a hydromedusa because it was raised from a hydroid in the laboratory in Bodega Bay).

Index

The single index incorporates all common and scientific names of phyla, classes, subclasses, orders, suborders, families, subfamilies, genera, species, and their phylogenetic subdivisions that are given in the list. A single entry is given for each common name; for example, sea strawberry is entered only as "strawberry, sea," and the order sea pens (Pennatulacea) appears only as "pens, sea." Each species is indexed by its specific name; thus, *Salacia articulata* is found only under *articulata, Salacia,* although the entry for *Salacia* directs the reader to the page on which entries for that genus begin.

Future of the Cnidaria and Ctenophora List

Proposed changes to this list are welcomed. The list will be revised as necessary, and subsequent editions will be published periodically. Readers who wish to recommend changes in listed scientific or common names, to reorganize the phyletic arrangement, or to add or delete species should (1) clearly identify the desired change or changes, (2) briefly and specifically justify the change with reference to literature sources, and (3) send the recommendation to the Chair, Committee on Common Names of Cnidaria and Ctenophora, c/o American Fisheries Society, 5410 Grosvenor Lane, Suite 110, Bethesda, Maryland 20814, USA, or directly to Stephen Cairns, Department of Invertebrate Zoology, Smithsonian Institution, Washington, D.C. 20560-0163. All suggested changes will be considered by the Committee for the next edition.

Acknowledgments

The Cnidaria and Ctenophora list was compiled by 11 systematists from four countries, all experts on their respective groups:

Dale R. Calder (Royal Ontario Museum, Toronto): Anthoathecatae (hydroids), Hydrina, Actinulida, Limnomedusae (in part), and Leptothecatae (hydroids);

Anita Brinckmann-Voss (Royal Ontario Museum, Toronto): Anthoathecatae (medusae), Limnomedusae (in part), Leptothecatae (medusae);

Clovis B. Castro (Museo Nacional, Rio de Janeiro): Octocorallia;

Daphne G. Fautin (University of Kansas, Lawrence): Actiniaria, Corallimorpharia, and Zoanthidea (with funding from the National Science Foundation, grant DEB 9978106 in the program Partnerships to Enhance Expertise in Taxonomy, or PEET);

Philip R. Pugh (Southampton Oceanography Centre, England): Siphonophora;

Stephen D. Cairns (Smithsonian Institution, Washington, D.C.): azooxanthellate Scleractinia and family Stylasteridae (Anthoathecatae).

Claudia E. Mills (Friday Harbor Laboratories): Cubozoa, Scyphozoa and Ctenophora (in part);

Walter C. Jaap (Florida Marine Research Institute, St. Petersburg): zooxanthellate Scleractinia;

Mary N. Arai (The University of Calgary, Alberta): Narcomedusae, Trachymedusae, Ceriantharia;

Stephen H. D. Haddock (Monterey Bay Aquarium): Ctenophora (in part);

Dennis M. Opresko (Oak Ridge National Laboratory, Tennessee): Antipatharia; and

As chair of the Committee, I thank my colleagues for their contributions, and we are all grateful to those who reviewed the draft lists at various stages of development. Without the help of those who spent so much time reviewing and checking these names, we could not have produced such a complete product. Acknowledgments and thanks are extended to F. M. Bayer, D. R. Calder, L. A. Gershwin, S. D. H. Haddock, E. Kelly, J. Lang, C. E. Mills, J. W. Porter, K. Sebens, W. Vervoort, S. T. Viada, and G. C. Williams. Deborah S. Lehman from the American Fisheries

Society's home office provided editorial assistance. Betsy Kulamer designed the figure layout.

The Committee has drawn freely on introductory material in the American Fisheries Society's most recent editions of *Common and Scientific Names of Fishes from the United States and Canada* (Robins et al. 1980, 1991) and from earlier volumes in the series *Common and Scientific Names of Aquatic Invertebrates from the United States and Canada: Mollusks* (Turgeon et al. 1998) and *Decapod Crustaceans* (Williams et al. 1989). Those lists served as models for this contribution.

STEPHEN D. CAIRNS
Washington, D.C.

References

CBE Style Manual Committee. 1983. CBE style manual, 5th edition. Council of Biology Editors, Inc., Bethesda, Maryland.

Grainger, E. H. 1971. Arctic zooplankton. Fisheries Research Board of Canada Bulletin 176. International Commission on Zoological Nomenclature. 1999. International code of zoological nomenclature, 4th edition. International Trust for Zoological Nomenclature, London.

Pritchard, D. W. 1955. Estuarine circulation patterns. Proceedings of the American Society of Civil Engineers 81(717):1–11.

Robins, C. R., R. M. Bailey, C. E. Bond, J. R. Brooker, E. A. Lachner, R. N. Lea, and W. B. Scott. 1980. A list of common and scientific names of fishes from the United States and Canada, 4th edition. American Fisheries Society, Special Publication 12, Bethesda, Maryland.

Robins, C. R., R. M. Bailey, C. E. Bond, J. R. Brooker, E. A. Lachner, R. N. Lea, and W. B. Scott. 1991. Common and scientific names of fishes from the United States and Canada, 5th edition. American Fisheries Society, Special Publication 20, Bethesda, Maryland.

Turgeon, D. D., J. F. Quinn, A. E. Bogan, E. V. Coan, F. G. Hochberg, W. G. Lyons, P. M. Mikkelsen, R. J. Neves, C. F. E. Roper, G. Rosenberg, B. Roth, A. Scheltema, F. G. Thompson, M. Vecchione, and J. D. Williams. 1998. Common and scientific names of aquatic invertebrates from the United States and Canada: mollusks (2nd edition). American Fisheries Society, Special Publication 26, Bethesda, Maryland.

Williams, A. B., L. G. Abele, D. L. Felder, H. H. Hobbs, Jr., R. B. Manning, P. A. McLaughlin, and I. Pérez Farfante. 1989. Common and scientific names of aquatic invertebrates from the United States and Canada: decapod crustaceans. American Fisheries Society, Special Publication 17, Bethesda, Maryland.

Bibliography

To serve as a guide for identification, terminology, and additional information on many of the species included in this list, a short bibliography is provided.

General

Austin, W. C. 1985. An annotated checklist of marine invertebrates in the cold temperate northeast Pacific, volume I. Khoyatan Marine Lab, Khoyatan, British Columbia.

Colin, P. I. 1978. Caribbean reef invertebrates and plants. T.F.H. Publications, Neptune City, New Jersey.

Kaplan, E. H. 1982. A field guide to coral reefs. Houghton Mifflin, Boston.

Sterrer, W. 1986. Marine fauna and flora of Bermuda. Wiley, New York.

Classes Cubozoa and Scyphozoa

Kramp, P. L. 1961. Synopsis of the medusae of the world. Journal of the Marine Biological Association of the United Kingdom 40.

Larson, R. J. 1976. Marine flora and fauna of the northeastern United States. Cnidaria: Scyphozoa. NOAA (National Oceanic and Atmospheric Administration) Technical Report NMFS (National Marine Fisheries Service) Circular 397.

Mayer, A. G. 1910. Medusae of the world. Scyphomedusae. III. Carnegie Institution, Washington, D.C.

Russell, F. S. 1970. The medusae of the British Isles. II. Pelagic Scyphozoa. Cambridge University Press, Cambridge, UK.

Orders Anthoathecatae and Leptothecatae (hydroids)

Cairns, S. D. 1986. A revision of the northwest Atlantic Stylasteridae. Smithsonian Contributions to Zoology 418.

Calder, D. R. 1970. Thecate hydroids from the shelf waters of northern Canada. Journal of the Fisheries Research Board of Canada 27:1501–1547.

Calder, D. R. 1972. Some athecate hydroids from the shelf waters of northern Canada. Journal of the Fisheries Research Board of Canada 29:217–228.

Calder, D. R. 1988. Shallow-water hydroids of Bermuda: the Athecatae. Royal Ontario Museum Life Sciences Contributions 148.

Calder, D. R. 1991. Shallow-water hydroids of Bermuda: the Thecatae, exclusive of Plumularioidea. Royal Ontario Museum Life Sciences Contributions 154.

Calder, D. R. 1997. Shallow-water hydroids of Bermuda: superfamily Plumularioidea. Royal Ontario Museum Life Sciences Contributions 161.

Cornelius, P. F. S. 1979. A revision of the species of Sertulariidae (Coelenterata: Hydroida) recorded from Britain and nearby seas. Bulletin of the British Museum (Natural History) Zoology 34:243–321.

Cornelius, P. F. S. 1982. Hydroids and medusae of the family Campanulariidae recorded from the eastern North Atlantic, with a world synopsis of genera. Bulletin of the British Museum (Natural History) Zoology 42:37–148.

Cornelius, P. F. S. 1995. North-West European thecate hydroids and their medusae. Parts 1 and 2. Synopses of the British Fauna, new series 50.

Fraser, C. M. 1937. Hydroids of the Pacific coast of Canada and the United States. University of Toronto Press, Toronto.

Fraser, C. M. 1944. Hydroids of the Atlantic coast of North America. University of Toronto Press, Toronto.

Orders Actinulida and Limnomedusae

Pennak, R. W. 1989. Fresh-water invertebrates of the United States, 3rd edition. Wiley, New York.

Orders Anthoathecatae and Leptothecatae (hydromedusae), Narcomedusae, and Trachymedusae

Arai, M. N., and A. Brinckmann-Voss. 1980. Hydromedusae of British Columbia and Puget Sound. Canadian Bulletin of Fisheries and Aquatic Sciences 204.

Burke, W. D. 1975. Pelagic Cnidaria of Mississippi Sound and adjacent waters. Gulf Research Reports 5:23–38.

Calder, D. R. 1971. Hydroids and hydromedusae of the southern Chesapeake Bay. Virginia Institute of Marine Science, special paper in marine science 1.

Kramp, P. L. 1942. Medusae. The "Godthaab" Expedition 1928. *Meddelelser om Grønland* 81(1).

Order Siphonophora

Alvariño, A. 1971. Siphonophores of the Pacific, with a review of the world distribution. Bulletin of the Scripps Institution of Oceanography of the University of California 16.

Bigelow, H. B. 1911. Reports on the scientific results of the expedition to the eastern tropical Pacific. XXIII: the Siphonophorae. Memoirs of the Museum of Comparative Zoology, Harvard 38:173–402.

Kirkpatrick, P. A., and P. R. Pugh. 1984. Siphonophores and velellids. Synopsis of the British Fauna, new series 29.

Totton, A. K. 1965. A synopsis of the Siphonophora. British Museum, London.

Order Antipatharia

Opresko, D. M. 1972. Redescriptions and reevaluations of the antipatharians described by L. F. de Pourtalès. Bulletin of Marine Science 22:950–1017.

Order Ceriantharia

Arai, M. N. 1965. A new species of *Pachycerianthus*, with a discussion of the genus and an appended glossary. Pacific Science 19:205–218.

Subclass Alcyonaria

Bayer, F. M. 1961. The shallow-water Octocorallia of the West Indian region. Studies on the Fauna of Curaçao and Other Caribbean Islands 12.

Bayer, F. M. 1981. Key to the genera of Octocorallia exclusive of the Pennatulacea, with diagnoses of new taxa. Proceedings of the Biological Society of Washington 94:902–947.

Orders Actiniaria and Corallimorpharia

Hand, C. 1955 (1954). The sea anemones of central California. Part 1. The corallimorpharian and athenarian anemones. Wasmann Journal of Biology 12:345–375.

Hand, C. 1955. The sea anemones of central California. Part 2. The endomyarian and mesomyarian anemones. Wasmann Journal of Biology 13:37–99.

Hand, C. 1956 (1955). The sea anemones of central California. Part 3. The acontiarian anemones. Wasmann Journal of Biology 13:189–251.

Order Zoanthidea

Herberts, C. 1987. *Ordre des Zoanthaires.* Pages 783–810 *in* P. P. Grassé, editor. Traité de zoologie, 3. Masson, Paris.

Walsh, G. E. 1967. An annotated bibliography of the family Zoanthidae, Epizoanthidae, and Parazoanthidae (Coelenterata, Zoanthidea). Institute of Marine Biology Technical Report 13, University of Hawaii, Kaneohe.

Order Scleractinia

Cairns, S. D. 1979. The deep-water Scleractinia of the Caribbean Sea and adjacent waters. Studies on the Fauna of Curaçao and Other Caribbean Islands 57(180).

Cairns, S. D. 1982. Stony corals of Carrie Bow Cay, Belize. Smithsonian Contributions to the Marine Sciences 12:271–302.

Roos, P. J. 1971. The shallow-water stony corals of the Netherlands Antilles. Studies on the Fauna of Curaçao and Other Caribbean Islands 37(130).

Smith, F. G. W. 1961. Atlantic reef corals. University of Miami Press, Coral Gables, Florida.

Wells, J. W. 1956. *Scleractinia.* Pages F328–444 *in* R. C. Moore, editor. Treatise on invertebrate paleontology. Part F: Coelenterata. Geological Society of America, New York.

Wells, J. W. 1973. New and old scleractinian corals from Jamaica. Bulletin of Marine Science 23:16–58.

Zlatarski, V. N. 1982. Les scléractiniaires de Cuba. Academie Bulgare des Sciences, Sofia, Bulgaria.

Phylum Ctenophora

Harbison, G. R. 1985. *On the classification and evolution of the Ctenophora.* Pages 70–100 *in* S. C. Morris, J. D. George, R. Gibson, and H. M. Platt, editors. The origins and relationships of lower invertebrates. Oxford University Press, Oxford, UK.

Harbison, G. R., and L. P. Madin. 1982. *The Ctenophora.* Pages 707–715 *in* S. P. Parker, editor. Synopsis and classification of living organisms. McGraw-Hill, New York.

Mills, C. E. 1987. *Phylum Ctenophora.* Pages 79–81 *in* E. N. Kozloff, editor. Marine invertebrates of the Pacific Northwest. University of Washington Press, Seattle.

PART I
Names of Cnidaria and Ctenophora

SCIENTIFIC NAME	OCCURRENCE	COMMON NAME

PHYLUM CNIDARIA (COELENTERATA)

Class Cubozoa—Sea wasps or box jellyfish

Carybdeidae

Carybdea alata Reynaud, 1830. . . . . A . . . .
* *Carybdea marsupialis* (Linnaeus, 1758) . . . . A-P . . . .
Tamoya haplonema Müller, 1859. . . . . A . . . .

Chirodropidae

Chiropsalmus quadrumanus (Müller, 1859) . . . . A . . . .

Class Scyphozoa—Jellyfish

ORDER STAUROMEDUSAE—STALKED JELLYFISH

SUBORDER ELEUTHEROCARPIDA

Lucernariidae

* *Haliclystus auricula* (Rathke, 1806). . . . . A . . . .
* *Haliclystus octoradiatus* (Lamarck, 1816) . . . . A-P . . . .
+ *Haliclystus salpinx* Clark, 1863. . . . . A-P . . . .
* *Haliclystus stejnegeri* Kishinouye, 1899 . . . . P . . . .
* *Haliclystus* sp. undescribed "*californiensis*". . . . . P . . . .
* *Haliclystus* sp. undescribed "*sanjuanensis*" . . . . P . . . .
Lucernaria quadricornis O. F. Müller, 1776. . . . . A . . . .
* Undescribed species "*Stenoscyphopsis vermiformis*". . . . . P . . . .

*Kyopodiidae

Kyopoda lamberti Larson, 1988 . . . . P . . . .

SUBORDER CLEISTOCARPIDA

Depastridae

*Thaumatoscyphinae

Manania atlantica (Berrill, 1962) . . . . A . . . .
* *Manania auricula* (Fabricius, 1780) . . . . A . . . .
Manania distincta (Kishinouye, 1910) . . . . P . . . .
Manania gwilliami Larson and Fautin, 1989 . . . . P . . . .
Manania handi Larson and Fautin, 1989 . . . . P . . . .

*Craterolophinae

* *Craterolophus convolvulus* (Johnston, 1835). . . . . A . . . .

ORDER CORONATAE—CROWN JELLYFISH

Atollidae

Atolla parva Russell, 1958 . . . . A-P. . . . .
Atolla tenella Hartlaub, 1909. . . . . Ac. . . . .

Occurrence abbreviations: A=Atlantic, P=Pacific, Ac=Arctic, E=Estuarine, F=Freshwater, I=Introduced.
*Change from first edition of 1991. +Not changed, but comment in Part II (Appendix 1).
^Species listed under both their hydroid and medusoid stages.

SCIENTIFIC NAME	OCCURRENCE	COMMON NAME
Atolla vanhoeffeni Russell, 1957	A-P	
Atolla wyvillei Haeckel, 1880	A-P	
***Atorellidae**		
* *Atorella octogonos* Mills, Larson and Youngbluth, 1987	A	
Atorella vanhoeffeni H. B. Bigelow, 1909	P	
Linuchidae		
* *Linuche unguiculata* (Swartz, 1788)	A	thimble jellyfish
Nausithoidae		
Nausithoe atlantica Broch, 1914	A	
Nausithoe punctata Kölliker, 1853	A	
Nausithoe rubra Vanhöffen, 1902	P	
Paraphyllinidae		
* *Paraphyllina intermedia* Mass, 1903	A-P	
Paraphyllina ransoni Russell, 1956	A-P	
Periphyllidae		
Periphylla periphylla (Péron and Lesueur, 1809)	A-P	
Periphyllopsis braueri Vanhöffen, 1902	A	
* *Periphyllopsis galathea* Kramp, 1959	A-P	
***Tetraplatidae**		
* *Tetraplatia volitans* Busch, 1951	A-P	
ORDER SEMAEOSTOMEAE		
Pelagiidae		
* *Chrysaora achlyos* Martin *et al.*, 1997	P	
Chrysaora fuscescens Brandt, 1835	P	
* *Chrysaora melanaster* Brandt, 1835	P-Ac	
Chrysaora quinquechirrha (Desor, 1848)	A	sea nettle
Pelagia colorata Russell, 1964	P	purple-striped jelly
Pelagia noctiluca (Forskål, 1775)	A	
Cyaneidae		
Cyanea capillata (Linnaeus, 1758)	A-P-Ac	lion's mane
* *Drymonema dalmatinum* Haeckel, 1880	A (I)	
***Ulmaridae**		
***Aureliinae**		
* *Aurelia aurita* (Linnaeus, 1758)	A-P(I)- Ac	moon jelly
* *Aurelia labiata* Chamisso and Eysenhardt, 1821	P	
* *Aurelia limbata* Brandt, 1835	P-Ac	brownbanded moon jelly
***Sthenoniinae**		
* *Phacellophora camtschatica* Brandt, 1835	A-P	fried egg jellyfish or egg yolk jelly
***Poraliinae**		
Poralia rufescens Vanhöffen, 1902	A-P	
***Stygiomedusinae**		
* *Stygiomedusa gigantea* (Browne, 1910)	P	
***Deepstariinae**		
Deepstaria enigmatica Russell, 1967	A-P	

Occurrence abbreviations: A=Atlantic, P=Pacific, Ac=Arctic, E=Estuarine, F=Freshwater, I=Introduced.
*Change from first edition of 1991. +Not changed, but comment in Part II (Appendix 1).
^Species listed under both their hydroid and medusoid stages.

SCIENTIFIC NAME	OCCURRENCE	COMMON NAME
ORDER RHIZOSTOMEAE		
Cassiopeidae—Upside-down jellyfish		
Cassiopea frondosa (Pallas, 1774)	A	
Cassiopea xamachana R. P. Bigelow, 1892	A	
Mastigiidae		
* *Phyllorhiza punctata* von Lendenfeld, 1884	A(I)-P(I)	Australian spotted jellyfish
Rhizostomatidae		
* *Rhopilema verrilli* (Fewkes, 1887)	A	mushroom jellyfish
* *Stomolophus meleagris* L. Agassiz, 1862	A-P	cannonball jellyfish or cabbagehead

Class Hydrozoa—Hydrozoans

*Subclass Leptolida

*ORDER ANTHOATHECATAE (=ATHECATAE)—ATHECATE HYDROIDS

SCIENTIFIC NAME	OCCURRENCE	COMMON NAME
+Clavidae		
+ *Clava multicornis* (Forsskål, 1775)	A-P	club hydroid
* *Cordylophora caspia* (Pallas, 1771)	A-P-E	freshwater hydroid
Corydendrium fruticosum Fraser, 1914	P	
Corydendrium parasiticum (Linnaeus, 1767)	A	
* *Merona cornucopiae* (Norman, 1864)	A-P	
* *Merona laxa* (Fraser, 1938)	P	
* *Rhizogeton ezoense* Yamada, 1964	P	
* *Rhizogeton fusiforme* L. Agassiz, 1862	A	
* *Rhizogeton nematophorum* Antsulevich, 1986	P	
* *Rhizogeton nudum* Broch, 1910	P-Ac	
Turritopsis fascicularis Fraser, 1943	A	
^* *Turritopsis nutricula* McCrady, 1857	A-P	
+Hydractiniidae		
Hydractinia aggregata Fraser, 1911	P	
^* *Hydractinia americana* (Edwards, 1972)	A	
^* *Hydractinia arge* (Clarke, 1882)	A	
Hydractinia armata Fraser, 1940	P	
Hydractinia californica Torrey, 1904	P	
+ *Hydractinia carica* Bergh, 1887	Ac	
Hydractinia carolinae Fraser, 1912	A	
* *Hydractinia echinata* (Fleming, 1828)	A	snailfur
^* *Hydractinia hooperii* (Sigerfoos, 1899)	A	
Hydractinia laevispina Fraser, 1922	P-Ac	
Hydractinia milleri Torrey, 1902	P	Miller hydractinia
Hydractinia monocarpa Allman, 1874	A	
Hydractinia polycarpa Fraser, 1938	P	
* *Hydractinia polyclina* L. Agassiz, 1862	A	
^* *Hydractinia selena* (Mills, 1976)	A	
* *Hydractinia symbiolongicarpus* Buss and Yund, 1989	A	
* *Hydractinia symbiopollicaris* Buss and Yund, 1989	A	
Hydractinia valens Fraser, 1941	A	
Janaria mirabilis Stechow, 1921	P	

SCIENTIFIC NAME	OCCURRENCE	COMMON NAME
Stylasteridae—hydrocorals		
Crypthelia glossopoma Cairns, 1986	A	
Distichopora foliacea Pourtalès, 1868	A	
Errinopora nanneca Fisher, 1938	P	
Errinopora pourtalesi (Dall, 1884)	P	
Pliobothrus symmetricus Pourtalès, 1868	A	
Stylantheca papillosa (Dall, 1884)	P	
Stylantheca petrograpta (Fisher, 1938)	P	
Stylantheca porphyra Fisher, 1931	P	
Stylaster californicus (Verrill, 1866)	P	California hydrocoral
Stylaster campylecus parageus (Fisher, 1938)	P	
Stylaster complanatus Pourtalès, 1867	A	
Stylaster duchassaingi Pourtalès, 1867	A	
* *Stylaster erubescens erubescens* Pourtalès, 1868	A	
Stylaster filogranus Pourtalès, 1871	A	
Stylaster laevigatus Cairns, 1986	A	
Stylaster miniatus (Pourtalès, 1868)	A	
Stylaster scabiosus Broch, 1935	P	
Stylaster stejnegeri (Fisher, 1938)	P	
Stylaster venustus (Verrill, 1870)	P	
Stylaster verrillii (Dall, 1884)	P	
***Rhysiidae**		
* *Rhysia fletcheri* Brinckmann-Voss, Lickey and Mills, 1993	P	
+Bougainvilliidae		
Aselomaris michaeli Berrill, 1948	A	
+ *Bimeria vestita* Wright, 1859	A-P	
^ *Bougainvillia carolinensis* (McCrady, 1859)	A	
Bougainvillia glorietta Torrey, 1904	P	
Bougainvillia inaequalis Fraser, 1944	A	
^* *Bougainvillia muscus* (Allman, 1863)	P	
^ *Bougainvillia rugosa* Clarke, 1882	A	
^+ *Bougainvillia superciliaris* (L. Agassiz, 1849)	A-P-Ac	
Dicoryne conferta (Alder, 1856)	A	
* *Dicoryne flexuosa* G. O. Sars, 1874	A	
Garveia annulata Nutting, 1901	P	orange hydroid
Garveia brevis (Fraser, 1918)	A	
+ *Garveia franciscana* (Torrey, 1902)	A-P-E	
Garveia gracilis (Clark, 1876)	P	
Garveia nutans Wright, 1859	P	
Garveia pusilla (Fraser, 1925)	P	
Garveia robusta (Torrey, 1902)	P	
Garveia tenella (Fraser, 1925)	P	
^ *Nemopsis bachei* L. Agassiz, 1849	A	
Rhizorhagium formosum (Fewkes, 1889)	P	
+ *Rhizorhagium roseum* M. Sars, 1874	A-P-Ac	
Pandeidae		
^+ *Amphinema dinema* (Péron and Lesueur, 1810)	A	
* *Hydrichthys mirus* Fewkes, 1887	A	
Ichthyocodium sarcotretis Jungersen, 1912	A	
*^ *Leuckartiara octona* (Fleming, 1823)	A	

Occurrence abbreviations: A=Atlantic, P=Pacific, Ac=Arctic, E=Estuarine, F=Freshwater, I=Introduced.
*Change from first edition of 1991. +Not changed, but comment in Part II (Appendix 1).
^Species listed under both their hydroid and medusoid stages.

SCIENTIFIC NAME	OCCURRENCE	COMMON NAME
Halimedusidae		
*^*Halimedusa typus* Bigelow, 1916	P	
Calycopsidae		
^*Bythotiara huntsmani* (Fraser, 1911)	P	
+Eudendriidae		
**Eudendrium album* Nutting, 1896	A-P	white stickhydroid
**Eudendrium arbuscula* Wright, 1859	Ac.	
Eudendrium attenuatum Allman, 1877	A-P	
Eudendrium breve Fraser, 1938	P	
Eudendrium californicum Torrey, 1902	P	California stickhydroid
+*Eudendrium capillare* Alder, 1856	A-P-Ac.	
Eudendrium carneum Clarke, 1882	A-P	red stickhydroid
Eudendrium certicaule Fraser, 1938	P	
Eudendrium cingulatum Stimpson, 1854	A	
Eudendrium cochleatum Allman, 1877	A-P	
Eudendrium dispar L. Agassiz, 1862	A	
Eudendrium exiguum Allman, 1877	A-P	
Eudendrium eximium Allman, 1877	A-P	
Eudendrium gracile Allman, 1877	A	
Eudendrium insigne Hincks, 1861	A-P	
Eudendrium irregulare Fraser, 1922	P	
Eudendrium laxum Allman, 1877	A	
Eudendrium rameum (Pallas, 1766)	A-P-Ac.	
Eudendrium ramosum (Linnaeus, 1758)	A-P	stickhydroid
Eudendrium rugosum Fraser, 1940	A	
+*Eudendrium vaginatum* Allman, 1863	A-P-Ac.	
+*Myrionema amboinense* Pictet, 1893	A	
+Monobrachiidae		
Monobrachium parasitum Mereschkowsky, 1877	A-P	
+Proboscidactylidae		
^*Proboscidactyla circumsabella* Hand, 1954	P	
^*Proboscidactyla flavicirrata* Brandt, 1835	P	
Proboscidactyla occidentalis (Fewkes, 1889)	P	
^*Proboscidactyla ornata* (McCrady, 1859)	A	
Moerisiidae		
^+*Moerisia lyonsi* Boulenger, 1908	A-P	
***Hydridae**		
Chlorohydra hadleyi Forrest, 1959	F	
Chlorohydra viridissima (Pallas, 1766)	F	green hydra
Hydra americana Hyman, 1929	F	white hydra
Hydra canadensis Rowan, 1930	F	Alberta hydra
Hydra carnea L. Agassiz, 1850	F	
Hydra cauliculata Hyman, 1938	F	
Hydra hymanae Hadley and Forrest, 1949	F	Jersey hydra
Hydra littoralis Hyman, 1931	F	swiftwater hydra
Hydra minima Forrest, 1963	F	
Hydra oligactis Pallas, 1766	F	brown hydra
Hydra oregona Griffin and Peters, 1939	F	Oregon hydra
Hydra pseudoligactis (Hyman, 1931)	F	false brown hydra

Occurrence abbreviations: A=Atlantic, P=Pacific, Ac=Arctic, E=Estuarine, F=Freshwater, I=Introduced.
*Change from first edition of 1991. +Not changed, but comment in Part II (Appendix 1).
^Species listed under both their hydroid and medusoid stages.

SCIENTIFIC NAME	OCCURRENCE	COMMON NAME
Hydra rutgersensis Forrest, 1963	F	
Hydra utahensis Hyman, 1931	F	Utah hydra
***Protohydridae**		
* *Protohydra leuckarti* Greeff, 1870	A-P-E	
Hydrocorynidae		
^ *Hydrocoryne bodegensis* J. T. Rees, Hand and Mills, 1976	P	
Cladocorynidae		
* *Cladocoryne floccosa* Rotch, 1871	A-P	
Zancleidae		
^* *Zanclea bomala* Boero, Bouillon, and Gravili, 2000	P	
^ *Zanclea costata* Gegenbaur, 1857	A-P	
* *Zanclea gemmosa* McCrady, 1859	A	
^* *Zanclella bryozoophila* Boero and Hewitt, 1992	P	
Milleporidae—fire corals		
Millepora alcicornis Linnaeus, 1758	A	fire coral
Millepora complanata Lamarck, 1816	A	bladed fire coral
Porpitidae		
Porpita porpita (Linnaeus, 1758)	A-P	blue button
Velella velella (Linnaeus, 1758)	A-P	by-the-wind sailor
Acaulidae		
Acaulis primarius Stimpson, 1854	A	
* *Hataia parva* Hirai and Yamada, 1965	P	
+Corymorphidae		
^* *Corymorpha bigelowi* (Maas, 1905)	P	
+ *Corymorpha groenlandica* (Allman, 1876)	Ac	
Corymorpha palma Torrey, 1902	P	fairypalm hydroid
+ *Corymorpha pendula* L. Agassiz, 1862	A	
+ *Euphysa farcta* (Miles, 1937)	A	
* *Euphysa peregrina* (Murbach, 1899)	A	
Euphysa ruthae Norenburg and Morse, 1983	P	
+Tubulariidae		
Ectopleura americana Petersen, 1990	A	
* *Ectopleura crocea* (L. Agassiz, 1862)	A-P	
^* *Ectopleura dumortierii* (van Beneden, 1844)	A	
Ectopleura grandis Fraser, 1944	A	
* *Ectopleura larynx* (Ellis and Solander, 1786)	A-P	ringed tubularia
* *Ectopleura marina* (Torrey, 1902)	P	
^ *Hybocodon prolifer* L. Agassiz, 1862	A-P-Ac	
Tubularia acadiae Petersen, 1990	A	
Tubularia aurea Fraser, 1936	P	
Tubularia harrimani Nutting, 1901	P	
+ *Tubularia indivisa* Linnaeus, 1758	A-P	tall tubularia
+ *Tubularia regalis* Boeck, 1860	P-Ac	
* *Zyzzyzus floridanus* Petersen, 1990	A	
Margelopsidae		
^* *Margelopsis gibbesii* (McCrady, 1859)	A	

Occurrence abbreviations: A=Atlantic, P=Pacific, Ac=Arctic, E=Estuarine, F=Freshwater, I=Introduced.
*Change from first edition of 1991. +Not changed, but comment in Part II (Appendix 1).
^Species listed under both their hydroid and medusoid stages.

SCIENTIFIC NAME	OCCURRENCE	COMMON NAME
+Candelabridae		
* *Candelabrum frichtmanni* Hewitt & Goddard, 2001	P	
+ *Candelabrum phrygium* (Fabricius, 1780)	A-Ac	
Monocoryne gigantea (Bonnevie, 1898)	Ac	
+Corynidae		
Coryne brachiata Nutting, 1901	P	
^* *Coryne cliffordi* (Brinckmann-Voss, 1989)	P	
Coryne corrugata Fraser, 1925	P	
Coryne crassa Fraser, 1914	P	
^* *Coryne eximia* Allman, 1859	A-P-Ac	
* *Coryne hincksii* Bonnevie, 1898	A-Ac	
^* *Coryne japonica* (Nagao, 1962)	P	
* *Coryne producta* (Wright, 1858)	P	
* *Coryne pusilla* Gaertner, 1774	A-Ac	
Dipurena bicircella J. T. Rees, 1977	P	
^ *Dipurena strangulata* McCrady, 1859	A	
^ *Sarsia apicula* (Murbach and Shearer, 1902)	P	
^* *Sarsia bella* Brinckmann-Voss, 2000	P	
Sarsia lovenii (M. Sars, 1846)	A	
^ *Sarsia occulta* Edwards, 1978	A	
^ *Sarsia princeps* (Haeckel, 1879)	P	
^+ *Sarsia tubulosa* (M. Sars, 1835)	A-P-Ac	clapper hydroid
***Sphaerocorynidae**		
^* *Sphaerocoryne agassizii* (McCrady, 1859)	A	
Cladonematidae		
^ *Cladonema californicum* Hyman, 1947	P	
Cladonema myersi Rees, 1949	P	
^+ *Cladonema radiatum* Dujardin, 1843	A-P	
Cladonema uchidai Hirai, 1958	P	
Solanderiidae		
Solanderia gracilis Duchassaing and Michelin, 1846	A	
***Pennariidae**		
^* *Pennaria disticha* Goldfuss, 1820	A	feather hydroid
***ORDER ANTHOATHECATAE (= ANTHOMEDUSAE)—HYDROMEDUSAE**		
Clavidae		
Oceania armata Kölliker, 1853	A	
^* *Turritopsis nutricula* (McCrady, 1857)	A(I)-P(I)	
+Hydractiniidae		
^* *Hydractinia americana* (Edwards, 1972)	A	
^* *Hydractinia arge* (Clarke, 1882)	A	
* *Hydractinia dubia* (Mayer, 1900)	A	
^* *Hydractinia hooperi* (Sigerfoos, 1899)	A	
* *Hydractinia minima* (Trinci, 1903)	A	
* *Hydractinia minuta* (Mayer, 1900)	A	
^* *Hydractinia selena* (Mills, 1976)	A	
Cytaeididae		
Cytaeis tetrastyla Eschscholtz, 1829	A	

Occurrence abbreviations: A=Atlantic, P=Pacific, Ac=Arctic, E=Estuarine, F=Freshwater, I=Introduced.
*Change from first edition of 1991. +Not changed, but comment in Part II (Appendix 1).
^Species listed under both their hydroid and medusoid stages.

SCIENTIFIC NAME	OCCURRENCE	COMMON NAME
Bougainvilliidae		
Bougainvillia britannica (Forbes, 1841)	A	
^ *Bougainvillia carolinensis* (McCrady, 1859)	A	
Bougainvillia frondosa Mayer, 1900	A	
^* *Bougainvillia muscus* (van Beneden, 1844)	A-P	
Bougainvillia niobe Mayer, 1894	A	
* *Bougainvillia principis* (Steenstrup, 1850)	A-P	
^ *Bougainvillia rugosa* Clarke, 1882	A	
^ *Bougainvillia superciliaris* (L.Agassiz, 1849)	A-P-Ac	
Koellikerina elegans (Mayer, 1900)	A	
Lizzia blondina Forbes, 1848	A	
Lizzia fulgurans (A. Agassiz, 1865)	A	
Lizzia gracilis (Mayer, 1900)	A	
^ *Nemopsis bachei* L. Agassiz, 1849	A	
Rathkeidae		
+ *Rathkea octopunctata* (M. Sars,1835)	A-P-Ac	
Pandeidae		
* *Amphinema australis* (Mayer, 1900)	A	
*^ *Amphinema dinema* (Péron and Lesueur, 1810)	A-P	
* *Amphinema platyhedos* Arai and Brinckmann-Voss, 1983	P	
Amphinema rugosum (Mayer, 1900)	A	
* *Amphinema turrida* (Mayer, 1900)	A-P	
* *Annatira affinis* (Hartlaub, 1913)	P	
Catablema multicirratum Kishinouye, 1910	P-Ac	
* *Catablema nodulosa* Bigelow, 1913	P	
Catablema vesicarium Hartlaub, 1914	A-Ac	constricted jellyfish
Cirrhitiara superba (Mayer, 1900)	A	
Eutiara mayeri Bigelow, 1918	A	
Geomackiea zephyrolata Mills, 1985	P	
Halitholus cirratus Hartlaub, 1913	A-Ac	
Halitholus pauper Hartlaub, 1913	A-P-Ac	
Halitiara formosa Fewkes, 1882	A	
Larsonia pterophylla (Haeckel, 1879)	A	
Leuckartiara nobilis Hartlaub, 1913	A-P	
^ *Leuckartiara octona* (Fleming, 1823)	A	
+ *Leuckartiara zacae* Bigelow, 1940	P	
* *Merga reesi* Russell, 1956	P	
Merga violacea (Agassiz and Mayer, 1899)	A	
Neoturris breviconis (Murbach and Shearer, 1902)	P-Ac	
+ *Neoturris fontata* (Bigelow, 1909)	P	
Niobia dendrotentacula Mayer, 1900	A	
+ *Pandea rubra* Bigelow, 1913	P	
Protiara haeckeli Hargitt, 1902	A	
+ *Stomotoca atra* L. Agassiz, 1862	P	
Halimedusidae		
+^ *Halimedusa typus* Bigelow, 1916	P	
***Calycopsidae**		
+ *Bythotiara depressa* Naumov, 1960	P	
^ *Bythotiara huntsmani* (Fraser, 1911)	P	
Bythotiara stilbosa Mills and Rees, 1979	P	

Occurrence abbreviations: A=Atlantic, P=Pacific, Ac=Arctic, E=Estuarine, F=Freshwater, I=Introduced.
*Change from first edition of 1991. +Not changed, but comment in Part II (Appendix 1).
^Species listed under both their hydroid and medusoid stages.

SCIENTIFIC NAME	OCCURRENCE	COMMON NAME
* *Calycopsis bigelowi* Vanhöffen, 1911	P	
Calycopsis nematophora Bigelow, 1913	P	
Calycopsis papillata Bigelow, 1918	A	
Calycopsis typa Fewkes, 1882	A	
* *Eumedusa birulai* (Linko, 1913)	Ac	
* *Heterotiara annonyma* Maas, 1905	P	
Proboscidactylidae		
^ *Proboscidactyla circumsabella* Hand, 1954	P	
^ *Proboscidactyla flavicirrata* Brandt, 1835	P	
^ *Proboscidactyla ornata* (McCrady, 1859)	A	
Moerisiidae		
^* *Moerisia lyonsi* Boulenger, 1908	A(E,I)-P	
Polyorchidae		
Polyorchis haplus Skogsberg, 1948	P	
* *Polyorchis penicillatus* (Eschscholtz, 1829)	P	penicillate jellyfish
Scrippsia pacifica Torrey, 1909	P	
***Zancleopsidae**		
* *Zancleopsis dichotoma* (Mayer, 1900)	A	
***Paragotoeidae**		
* *Paragotoea bathybia* Kramp, 1942	P	
Hydrocorynidae		
^ *Hydrocoryne bodegensis* J. T. Rees, Hand and Mills, 1976	P	
***Family incertae sedis (suborder Sphaerocorynida)**		
Euphysilla peterseni Allwein, 1967	P	
Zancleidae		
^* *Zanclea bomala* Boero, Bouillon and Gravili, 2000	P	
^* *Zanclea costata* Gegenbaur, 1857	A	
* *Zanclea gemmosa* McCrady, 1859	A	
^* *Zanclella bryozoophila* Boero and Hewitt, 1992	P	
+Corymorphidae		
* *Euphysa aurata* Forbes, 1848	A	
Euphysa flammea (Linko, 1905)	A-Ac	
Euphysa japonica (Maas, 1909)	P	
Euphysa tentaculata Linko, 1905	A-P	
* *Euphysa vervoorti* Brinckmann-Voss and Arai, 1998	P	
^* *Corymorpha bigelowi* Maas, 1905	P	
* *Corymorpha forbesi* (Mayer, 1894)	A	
* *Corymorpha gracilis* (Brooks, 1882)	A	
* *Corymorpha nutans* M. Sars, 1835	A-P	
***Tubulariidae**		
^* *Ectopleura dumortierii* (van Beneden, 1844)	A	
Ectopleura pacifica Mayer, 1900	A	
Hybocodon pendulus (L.Agassiz, 1862)	A	one armed jellyfish
^ *Hybocodon prolifer* L. Agassiz, 1862	A-P-Ac	
***Margelopsidae**		
^* *Margelopsis gibbesii* (McCrady, 1859)	A	

Occurrence abbreviations: A=Atlantic, P=Pacific, Ac=Arctic, E=Estuarine, F=Freshwater, I=Introduced.
*Change from first edition of 1991. +Not changed, but comment in Part II (Appendix 1).
^Species listed under both their hydroid and medusoid stages.

SCIENTIFIC NAME	OCCURRENCE	COMMON NAME
***Corynidae**		
^* *Coryne cliffordi* (Brinckmann-Voss, 1989)	P	
^* *Coryne eximia* Allman, 1859	P	
^* *Coryne japonica* (Nagao, 1962)	P	
+ *Dicodonium floridana* Mayer, 1900	A	
Dicodonium jeffersoni (Mayer, 1910)	A	
Dipurena halterata (Forbes, 1846)	A	
Dipurena ophiogaster Haeckel, 1879	P	
^ *Dipurena strangulata* McCrady, 1859	A	
^ *Sarsia apicula* (Murbach and Shearer, 1902)	P	
^* *Sarsia bella* Brinckmann-Voss, 2000	P	
Sarsia gemmifera Forbes, 1848	A	
Sarsia hargitti Mayer, 1910	A	
^* *Sarsia occulta* Edwards, 1978	A	
^ *Sarsia princeps* (Haeckel, 1879)	A-P-Ac	
^ *Sarsia tubulosa* (M. Sars, 1835)	A-P-Ac	clapper medusa
***Sphaerocorynidae**		
*^ *Sphaerocoryne agassizii* (McCrady, 1859)	A	
Cladonematidae		
^ *Cladonema californicum* Hyman, 1947	P	
^* *Cladonema radiatum* Dujardin, 1843	A(I)-P(I)	
***Pennariidae**		
^* *Pennaria disticha* Goldfuss, 1820	P	
***Family incertae sedis (suborder Tubulariida)**		
* *Plotocnide borealis* Wagner, 1885	P	
***Family incertae sedis (order Anthoathecatae-Anthomedusae)**		
* *Paulinum lineatum* Brinkmann-Voss and Arai, 1998	P	
***Trichydridae**		
* *Trichydra pudica* Wright, 1858	P	
ORDER LIMNOMEDUSAE		
Olindiidae		
Aglauropsis aeora Mills, J. T. Rees and Hand, 1976	P	
Calpasoma dactylopterum Fuhrmann, 1939	F	
* *Craspedacusta sowerbyi* Lankester, 1880	F, I	freshwater jellyfish
Cubaia aphrodite Mayer, 1894	A	
Eperemetus typus H.B. Bigelow, 1915	P	
Gonionemus vertens L. Agassiz, 1862	A-P	clinging jellyfish
Gossea brachymera H.B. Bigelow, 1909	A	
* *Maeotias marginata* (Modeer, 1791)	A(E,I)-P (E,I)	
Olindias tenuis (Fewkes, 1882)	A	
Vallentinia adherens Hyman, 1947	P	
***ORDER LEPTOTHECATAE (=THECATAE)—THECATE HYDROIDS**		
***Melicertidae**		
^* *Melicertum octocostatum* (M. Sars, 1835)	A	

Occurrence abbreviations: A=Atlantic, P=Pacific, Ac=Arctic, E=Estuarine, F=Freshwater, I=Introduced.
*Change from first edition of 1991. +Not changed, but comment in Part II (Appendix 1).
^Species listed under both their hydroid and medusoid stages.

SCIENTIFIC NAME	OCCURRENCE	COMMON NAME
Aequoreidae		
^* *Aequorea forskalea* Péron and Lesueur, 1810	A	
^ *Aequorea victoria* (Murbach and Shearer, 1902)	P	
Blackfordiidae		
^* *Blackfordia virginica* Mayer, 1910	A-P-E	
Phialellidae		
Opercularella lacerata (Johnston, 1847)	A-P-Ac	
Opercularella pumila Clark, 1875	A	
* *Opercularella rugosa* (Nutting, 1901)	P	
^ *Phialella fragilis* (Uchida, 1938)	P	
^ *Phialella zappai* Boero, 1987	P	
Calycellidae		
Calycella gracilis Hartlaub, 1897	Ac	
+ *Calycella syringa* (Linnaeus, 1767)	A-P-Ac	creeping bell hydroid
Tetrapoma quadridentatum (Hincks, 1874)	A-Ac	
+Family incertae sedis		
Cuspidella costata Hincks, 1868	A	
Cuspidella grandis Hincks, 1868	A-P	
Cuspidella humilis (Alder, 1862)	A-P-Ac	
Cuspidella procumbens Kramp, 1911	Ac	
Egmundella fasciculata Fraser, 1940	A	
Egmundella gracilis Stechow, 1921	P	
Keratosum maximum (Levinsen, 1893)	A-Ac	
***Laodiceidae**		
^ *Ptychogena lactea* A. Agassiz, 1865	A-P-Ac	
***Tiarannidae**		
+ *Modeeria rotunda* (Quoy and Gaimard, 1827)	A-P	
Stegopoma plicatile (M. Sars, 1863)	A-P-Ac	
***Eucheilotidae**		
^+ *Eucheilota bakeri* (Torrey, 1904)	P	
Lovenellidae		
^ *Lovenella gracilis* (Clarke, 1882)	A	
Lovenella grandis Nutting, 1901	A	
* *Lovenella nodosa* Fraser, 1938	P	
* *Lovenella producta* (G. O. Sars, 1874)	A-P	
Lovenella rugosa Fraser, 1948	P	
Haleciidae		
Halecium annulatum Torrey, 1902	P	
+ *Halecium articulosum* Clark, 1875	A-P	
Halecium beanii (Johnston, 1838)	A-P	
Halecium bermudense Congdon, 1907	A	
Halecium capillare (Pourtalès, 1869)	A	
Halecium corrugatum Nutting, 1899	A-P	
Halecium curvicaule von Lorenz, 1886	A-Ac	
* *Halecium delicatulum* Coughtrey, 1876	A-P	
Halecium densum Calkins, 1899	P	
Halecium diminutivum Fraser, 1940	A	

Occurrence abbreviations: A=Atlantic, P=Pacific, Ac=Arctic, E=Estuarine, F=Freshwater, I=Introduced.
*Change from first edition of 1991. +Not changed, but comment in Part II (Appendix 1).
^Species listed under both their hydroid and medusoid stages.

SCIENTIFIC NAME	OCCURRENCE	COMMON NAME
Halecium fruticosum Fraser, 1943	A	
Halecium groenlandicum Kramp, 1911	A-Ac	
+*Halecium halecinum* (Linnaeus, 1758)	A-P	herringbone hydroid
Halecium insolens Fraser, 1938	P	
Halecium kofoidi Torrey, 1902	P	
Halecium labrosum Alder, 1859	A-P-Ac	
Halecium macrocephalum Allman, 1877	A-P	
Halecium minutum Broch, 1903	A-Ac	
Halecium muricatum (Ellis and Solander, 1786)	A-P-Ac	sea hedgehog hydroid
Halecium nanum Alder, 1859	A-P	
Halecium ornatum Nutting, 1901	P	
Halecium paucinodum (Fraser, 1947)	P	
Halecium pygmaeum Fraser, 1911	P	
Halecium reversum Nutting, 1901	P	
* *Halecium scutum* Clark, 1877	A-P-Ac	
* *Halecium sessile* Norman, 1867	A	
Halecium speciosum Nutting, 1901	P-Ac	
Halecium telescopicum Allman, 1888	P	
Halecium tenellum Hincks, 1861	A-P-Ac	
* *Halecium undulatum* Billard, 1921	A-Ac	
Halecium vagans Fraser, 1938	P	
Halecium washingtoni Nutting, 1901	P	
Halecium wilsoni Calkins, 1899	P	
Hydrodendron alternatum (Fraser, 1938)	P	
Hydrodendron carchesium (Fraser, 1914)	P	
Hydrodendron corrugatum (Fraser, 1936)	P	
Hydrodendron expansum (Fraser, 1948)	P	
Hydrodendron gracile (Fraser, 1914)	A-P	
+ *Hydrodendron mirabile* (Hincks, 1866)	A	
Mitrocomium medusiferum Torrey, 1902	P	
* *Sagamihydra dyssymetra* (Billard, 1929)	A	
+Kirchenpaueriidae		
* *Kirchenpaueria paucinema* (Fraser, 1940)	P	
* *Kirchenpaueria plumularioides* (Clark, 1877)	P	
* *Pycnotheca allmani* (Torrey, 1904)	P	
+ *Ventromma halecioides* (Alder, 1859)	A	
+Plumulariidae		
Callicarpa chazaliei Versluys, 1899	A-P	plumed hydroid
* *Dentitheca dendritica* (Nutting, 1900)	A	
* *Hippurella annulata* Allman, 1877	A	
* *Monotheca margaretta* Nutting, 1900	A-P	
Nemertesia americana (Nutting, 1900)	A	
* *Nemertesia antennina* (Linnaeus, 1758)	A	sea beard
+ *Nemertesia irregularis* (Fraser, 1938)	P	
Nemeresia pinnata (Nutting, 1900)	A	
Nemertesia polynema (Fraser, 1948)	P	
Nemertesia rugosa (Nutting, 1900)	A	
Nemertesia simplex (Allman, 1877)	A	
Nemertesia tetraseriata (Fraser, 1938)	P	
Nemertesia verticillata (Fraser, 1925)	P	
Plumularia attenuata Allman, 1877	A-P	

Occurrence abbreviations: A=Atlantic, P=Pacific, Ac=Arctic, E=Estuarine, F=Freshwater, I=Introduced.
*Change from first edition of 1991. +Not changed, but comment in Part II (Appendix 1).
^Species listed under both their hydroid and medusoid stages.

SCIENTIFIC NAME	OCCURRENCE	COMMON NAME
Plumularia constricta (Fraser, 1948)	P	
Plumularia exilis Fraser, 1948	P	
Plumularia filicula Allman, 1877	A-P	
+ *Plumularia floridana* Nutting, 1900	A-P	
Plumularia goodei Nutting, 1900	P	
Plumularia gracilis (Fraser, 1948)	P	
Plumularia gracillima G. O. Sars, 1873	A	
Plumularia insolens Fraser, 1948	P	
Plumularia integra Fraser, 1948	P	
Plumularia inverta (Fraser, 1948)	P	
Plumularia irregularis Fraser, 1948	P	
Plumularia lagenifera Allman, 1885	P	
Plumularia megalocephala Allman, 1877	A-P	
Plumularia meganema Fraser, 1948	P	
Plumularia mobilis Fraser, 1948	P	
Plumularia multiramosa Fraser, 1948	P	
Plumularia parva Fraser, 1948	P	
Plumularia reversa Fraser, 1948	P	
* *Plumularia septata* Fraser, 1938	P	
+ *Plumularia setacea* (Linnaeus, 1758)	A-P	little seabristle
Plumularia sinuosa Fraser, 1938	P	
+ *Plumularia strictocarpa* Pictet, 1893	A	
Plumularia virginiae Nutting, 1900	P	
+Halopterididae		
Antennella avalonia Torrey, 1902	P	
Antennella quadriaurita Ritchie, 1909	A	
* *Antennella secundaria* (Gmelin, 1791)	A-P	
* *Halopteris alternata* (Nutting, 1900)	A-P	
Halopteris carinata Allman, 1877	A	
Halopteris catharina (Johnston, 1833)	A	
+ *Halopteris clarkei* (Nutting, 1900)	A	
Halopteris geminata (Allman, 1877)	A	
* *Halopteris tenella* (Verrill, 1874)	A-P	
Monostaechas quadridens (McCrady, 1859)	A-P	
* *Nuditheca dallii* (Clark, 1877)	P	
+Aglaopheniidae		
+ *Aglaophenia apocarpa* Allman, 1877	A-P	
Aglaophenia constricta Allman, 1877	A	
Aglaophenia diegensis Torrey, 1904	P	
Aglaophenia dispar Fraser, 1948	P	
Aglaophenia diversidentata Fraser, 1948	P	
* *Aglaophenia dubia* Nutting, 1900	A	
Aglaophenia epizoica Fraser, 1948	P	
Aglaophenia fluxa Fraser, 1948	P	
Aglaophenia inconspicua Torrey, 1904	P	
Aglaophenia integriseptata Fraser, 1948	P	
+ *Aglaophenia latecarinata* Allman, 1877	A	
Aglaophenia lateseptata Fraser, 1948	P	
Aglaophenia latirostris Nutting, 1900	P	
Aglaophenia octocarpa Nutting, 1900	P	
Aglaophenia pinguis Fraser, 1938	P	

Occurrence abbreviations: A=Atlantic, P=Pacific, Ac=Arctic, E=Estuarine, F=Freshwater, I=Introduced.
*Change from first edition of 1991. +Not changed, but comment in Part II (Appendix 1).
^Species listed under both their hydroid and medusoid stages.

SCIENTIFIC NAME	OCCURRENCE	COMMON NAME
* *Aglaophenia pluma* (Linnaeus, 1758)	P	podded hydroid
Aglaophenia prominens Fraser, 1938	P	
+ *Aglaophenia rhynchocarpa* Allman, 1877	A	
Aglaophenia struthionides (Murray, 1860)	P	ostrichplume hydroid
Aglaophenia tridentata Versluys, 1999	A	
+ *Aglaophenia trifida* L. Agassiz, 1862	A	
Cladocarpus dolichotheca Allman, 1877	A	
Cladocarpus flexilis Verrill, 1885	A	
Cladocarpus gracilis Fraser, 1948	P	
+ *Cladocarpus integer* (G. O. Sars, 1874)	A	
Cladocarpus moderatus Fraser, 1948	P	
Cladocarpus obliquus Nutting, 1900	A	
Cladocarpus paradiseus Allman, 1877	A	
Cladocarpus pinguis Fraser, 1948	P	
Cladocarpus septatus Nutting, 1900	A	
Cladocarpus sigma (Allman, 1877)	A	
Cladocarpus tenuis Clarke, 1879	A	
Cladocarpus vancouverensis Fraser, 1914	P	
Cladocarpus ventricosus Allman, 1877	A	
Gymnangium sinuosum (Fraser, 1925)	A	
Gymnangium speciosum (Allman, 1877)	A	
Lytocarpia bispinosa (Allman, 1877)	A	
Lytocarpia myriophyllum (Linnaeus, 1758)	A	pheasant-tail hydroid
+ *Macrorhynchia allmani* (Nutting, 1900)	A	
* *Macrorhynchia philippina* Kirchenpauer, 1872	A-P	white stinger
***Lafoeidae**		
Acryptolaria abies (Allman, 1877)	A	
Acryptolaria conferta (Allman, 1877)	A-P	
Acryptolaria longitheca (Allman, 1877)	A	
Acryptolaria pulchella (Allman, 1888)	P	
Acryptolaria rectangularis (Jarvis, 1922)	A	
Acryptolaria triserialis (Fraser, 1913)	A	
Cryptolaria pectinata (Allman, 1888)	A	
Filellum serpens (Hassall, 1848)	A-P-Ac	
Filellum serratum (Clarke, 1879)	A	
* *Grammaria abietina* (M. Sars, 1850)	A-P-Ac	
Grammaria gracilis Stimpson, 1854	A	
+ *Grammaria immersa* Nutting, 1901	P-Ac	
Lafoea adhaerens Nutting, 1901	P	
Lafoea adnata Fraser, 1925	P	
Lafoea coalescens Allman, 1877	A	
* *Lafoea dumosa* (Fleming, 1820)	A-P-Ac	
* *Lafoea fruticosa* (M. Sars, 1850)	A-P-Ac	
* *Lafoea gracillima* (Alder, 1856)	A-P-Ac	
* *Lafoea tenellula* Allman, 1877	A	
Hebella venusta (Allman, 1877)	A	
* *Hebellopsis cylindrica* (von Lendenfeld, 1885)	A-P	
Hebellopsis expansa (Fraser, 1938)	P	
Hebellopsis scandens (Bale, 1888)	A-P	
* *Scandia corrugata* Fraser, 1938	P	
Scandia mutabilis (Ritchie, 1907)	A-P	

Occurrence abbreviations: A=Atlantic, P=Pacific, Ac=Arctic, E=Estuarine, F=Freshwater, I=Introduced.
*Change from first edition of 1991. +Not changed, but comment in Part II (Appendix 1).
^Species listed under both their hydroid and medusoid stages.

SCIENTIFIC NAME	OCCURRENCE	COMMON NAME
Zygophylax adhaerens Fraser, 1938	P	
* *Zygophylax carolinus* (Fraser, 1911)	P	
* *Zygophylax cervicornis* (Nutting, 1905)	P	
* *Zygophylax convallarius* (Allman, 1877)	A	
* *Zygophylax crassitheca* (Fraser, 1941)	A	
* *Zygophylax pinnatus* (G. O. Sars, 1874)	A	
* *Zygophylax reflexus* (Fraser, 1948)	P	
+ *Zygophylax rigidus* (Fraser, 1948)	P	
+ *Zygophylax robustus* (Verrill, 1873)	A-P	
+Campanulariidae		
Campanularia castellata Fraser, 1925	P	
+ *Campanularia emarginata* Fraser, 1938	P	
Campanularia groenlandica Levinsen, 1893	A-P-Ac.	
Campanularia hincksii Alder, 1856	A-P	
Campanularia macroscypha Allman, 1877	A	
Campanularia ritteri Nutting, 1901	P	
+ *Campanularia volubilis* (Linnaeus, 1758)	A-P-Ac.	
Clytia attenuata (Calkins, 1899)	P	
* *Clytia denticulata* (Clark, 1877)	P	
+ *Clytia exilis* Fraser, 1948	P	
Clytia fascicularis Fraser, 1938	P	
* *Clytia gracilis* (M. Sars, 1850)	A-P	
^ *Clytia gregaria* (L. Agassiz, 1862)	P	
^* *Clytia hemisphaerica* (Linnaeus, 1767)	A-P	
* *Clytia hendersoni* Torrey, 1904	P	
Clytia hesperia (Torrey, 1904)	P	
Clytia irregularis Fraser, 1938	P	
Clytia kincaidi (Nutting, 1899)	A-P	
+ *Clytia linearis* (Thornely, 1900)	A-P	
^ *Clytia lomae* (Torrey, 1909)	P	
Clytia macrotheca (Perkins, 1908)	A	
^* *Clytia noliformis* auct.	A	
* *Clytia paulensis* (Vanhöffen, 1910)	A-P	
Clytia universitatis Torrey, 1904	P	
* *Gonothyraea loveni* (Allman, 1859)	A-P-Ac.	
Hartlaubella gelatinosa (Pallas, 1766)	A-P-Ac.	
Laomedea altitheca (Fraser, 1948)	P	
Laomedea amphora L. Agassiz, 1862	A	
Laomedea calceolifera (Hincks, 1871)	A	
Laomedea exigua M. Sars, 1857	P	
Laomedea flexuosa Alder, 1857	A	
Laomedea inornata (Nutting, 1901)	P	
+ *Laomedea neglecta* Alder, 1856	A	
+ *Obelia bidentata* Clark, 1875	A-P	doubletoothed hydroid
^+ *Obelia dichotoma* (Linnaeus, 1758)	A-P	sea thread hydroid
Obelia geniculata (Linnaeus, 1758)	A-P-Ac	knotted thread hydroid
^+ *Obelia longissima* (Pallas, 1766)	A-P-Ac.	
Obelia multidentata Fraser, 1914	P	
Obelia plicata Hincks, 1868	A-P-Ac.	
* *Obelia racemosa* Fraser, 1941	A	
^* *Orthopyxis compressa* (Clark, 1877)	P	

Occurrence abbreviations: A=Atlantic, P=Pacific, Ac=Arctic, E=Estuarine, F=Freshwater, I=Introduced.
*Change from first edition of 1991. +Not changed, but comment in Part II (Appendix 1).
^Species listed under both their hydroid and medusoid stages.

SCIENTIFIC NAME	OCCURRENCE	COMMON NAME
Orthopyxis everta (Clark, 1876)	P	
+^*Orthopyxis integra* (Macgillivray, 1842)	A-P-Ac.	
Orthopyxis minor (Fraser, 1938)	P	
Orthopyxis sargassicola (Nutting, 1915)	A	
Rhizocaulus verticillatus (Linnaeus, 1758)	A-P-Ac	horsetail hydroid
* *Tulpa crenata* Allman, 1876	A-P-Ac.	
Bonneviellidae		
Bonneviella regia (Nutting, 1901)	P	
Thyroscyphidae		
* *Cnidoscyphus marginatus* (Allman, 1877)	A	
Thyroscyphus ramosus Allman, 1877	A	
+Syntheciidae		
+ *Hincksella cylindrica* (Bale, 1888)	A-P	
* *Hincksella projecta* (Fraser, 1938)	P	
Synthecium symmetricum Fraser, 1938	P	
+ *Synthecium tubithecum* (Allman, 1877)	A	
+Sertulariidae		
Abietinaria abietina (Linnaeus, 1758)	A-P-Ac.	sea fir
Abietinaria alexanderi Nutting, 1904	P	
Abietinaria amphora Nutting, 1904	P	
+ *Abietinaria anguina* (Trask, 1857)	P	
Abietinaria annulata (Kirchenpauer, 1884)	P	
+ *Abietinaria costata* (Nutting, 1901)	P	
Abietinaria expansa Fraser, 1938	P	
Abietinaria filicula (Ellis and Solander, 1786)	A-P	fern hydroid
* *Abietinaria gigantea* (Clark, 1877)	P	
Abietinaria gracilis Nutting, 1904	P	
Abietinaria greenei (Murray, 1860)	P	
* *Abietinaria inconstans* (Clark, 1877)	P	
* *Abietinaria kincaidi* (Nutting, 1901)	A-P	
Abietinaria pacifica Stechow, 1923	P	
* *Abietinaria pulchra* (Nutting, 1904)	A-P-Ac.	
Abietinaria rigida Fraser, 1911	P	
* *Abietinaria thuiarioides* (Clark, 1877)	A-P	
Abietinaria traski (Torrey, 1902)	P	
* *Abietinaria turgida* (Clark, 1877)	A-P-Ac.	
* *Abietinaria variabilis* (Clark, 1877)	P	
Amphisbetia erecta (Fraser, 1938)	P	
Amphisbetia furcata (Trask, 1857)	P	
Amphisbetia kurilae (Nutting, 1904)	P	
* *Amphisbetia operculata* (Linnaeus, 1758)	P	
Diphasia corniculata (Murray, 1860)	P	
Diphasia digitalis (Busk, 1852)	A	
Diphasia fallax (Johnston, 1847)	A	
+ *Diphasia pinastrum* (Cuvier, 1830)	A	sea pine hydroid
Diphasia rosacea (Linnaeus, 1758)	A	lily hydroid
Diphasia tropica Nutting, 1904	A	
* *Dynamena crisioides* Lamouroux, 1824	A	
* *Dynamena dalmasi* (Versluys, 1899)	A	
Dynamena dispar (Fraser, 1938)	P	

Occurrence abbreviations: A=Atlantic, P=Pacific, Ac=Arctic, E=Estuarine, F=Freshwater, I=Introduced.
*Change from first edition of 1991. +Not changed, but comment in Part II (Appendix 1).
^Species listed under both their hydroid and medusoid stages.

SCIENTIFIC NAME	OCCURRENCE	COMMON NAME
+*Dynamena disticha* (Bosc, 1802)	A-P	
+*Dynamena pumila* (Linnaeus, 1758)	A	sea oak
Dynamena quadridentata (Ellis and Solander, 1786)	A	
Dynamena tropica Stechow, 1926	A	
**Fraseroscyphus sinuosus* Boero and Bouillon, 1993	P	
Hydrallmania distans Nutting, 1899	P	
Hydrallmania falcata (Linnaeus, 1758)	A	sickle hydroid
Hydrallmania franciscana (Trask, 1857)	P	
Idiellana pristis (Lamouroux, 1816)	A	
**Pericladium mirabilis* (Verrill, 1873)	A-P-Ac.	
**Pericladium trilateralis* (Fraser, 1936)	P	
Selaginopsis cedrina (Linnaeus, 1758)	P	
**Selaginopsis cylindrica* (Clark, 1877)	P	
Selaginopsis hartlaubi (Nutting, 1904)	P	
Selaginopsis obsoleta (Lepechin, 1778)	P	
Selaginopsis pinaster (Lepechin, 1783)	P	
Selaginopsis triserialis Mereschkowsky, 1878	P	
Sertularella albida Kirchenpauer, 1884	P	
Sertularella areyi Nutting, 1904	A	
**Sertularella clarkii* Mereschkowsky, 1878	P	
Sertularella complexa Nutting, 1904	P	
**Sertularella conella* Stechow, 1920	P	
+*Sertularella conica* Allman, 1877	A	
+*Sertularella diaphana* (Allman, 1885)	A	
**Sertularella ellisi* (Deshayes and Milne Edwards, 1836)	A-P-Ac.	
**Sertularella flabella* (Nutting, 1904)	P	
Sertularella gayi (Lamouroux, 1821)	A	
Sertularella gigantea Mereschowsky, 1878	A	
Sertularella polyzonias (Linnaeus, 1758)	A-P-Ac	great tooth hydroid
Sertularella rugosa (Linnaeus, 1758)	A-P	snail trefoil hydroid
Sertularella tanneri Nutting, 1904	P	
**Sertularella tenella* (Alder, 1856)	A-P-Ac.	
**Sertularia argentea* Linnaeus, 1758	A-P	squirrel's tail hydroid
+*Sertularia carolinensis* Verrill, 1872	A	
**Sertularia cupressina* Linnaeus, 1758	A.	sea cypress hydroid
**Sertularia dalli* (Nutting, 1904)	A-P	
Sertularia elegans (Kirchenpauer, 1884)	P	
Sertularia fabricii Levinsen, 1893	A-P	
Sertularia latiuscula Stimpson, 1854	A	
**Sertularia plumosa* (Clark, 1877)	P-Ac.	
+*Sertularia plumulifera* (Allman, 1877)	A	
**Sertularia robusta* (Clark, 1877)	A-P-Ac.	
Sertularia schmidti Kudelin, 1914	A-Ac	
**Sertularia similis* Clark, 1877	A-P-Ac.	
Sertularia tenera G. O. Sars, 1874	A-P-Ac.	
**Symplectoscyphus amphoriferus* (Allman, 1877)	A	
Symplectoscyphus elegans (Nutting, 1904)	P	
Symplectoscyphus levinseni (Nutting, 1904)	P	
Symplectoscyphus pedrensis (Torrey, 1904)	P	
**Symplectoscyphus pinnatus* (Clark, 1877)	P-Ac.	
Symplectoscyphus sinuosus (Fraser, 1948)	P	
Symplectoscyphus tricuspidatus (Alder, 1856)	A-P-Ac.	

Occurrence abbreviations: A=Atlantic, P=Pacific, Ac=Arctic, E=Estuarine, F=Freshwater, I=Introduced.
*Change from first edition of 1991. +Not changed, but comment in Part II (Appendix 1).
^Species listed under both their hydroid and medusoid stages.

SCIENTIFIC NAME	OCCURRENCE	COMMON NAME
Symplectoscyphus turgidus (Trask, 1857)	P	
Tamarisca tamarisca (Linnaeus, 1758)	A	sea tamarisk
* *Thuiaria alba* Fraser, 1911	P	
* *Thuiaria alternitheca* Levinsen, 1893	P	
* *Thuiaria articulata* (Pallas, 1766)	A-P	sea spleenwort
* *Thuiaria carica* Levinsen, 1893	A-P-Ac	
* *Thuiaria desmoides* (Torrey, 1902)	P	
* *Thuiaria fraseri* (Calder, 1991)	P	
* *Thuiaria laxa* Allman, 1874	A-Ac	
Thuiaria thuja (Linnaeus, 1758)	A-P-Ac	bottlebrush hydroid
* *Tridentata distans* (Lamouroux, 1816)	A-P	
Tridentata flowersi (Nutting, 1904)	A	
+ *Tridentata marginata* (Kirchenpauer, 1864)	A-P	
Tridentata tumida (Allman, 1877)	A	
Tridentata turbinata (Lamouroux, 1816)	A	

+ORDER ACTINULIDA

Halammohydridae

SCIENTIFIC NAME	OCCURRENCE	COMMON NAME
Halammohydra schulzei Remane, 1927	A	

*ORDER LEPTOTHECATAE (=LEPTOMEDUSAE)—HYDROMEDUSAE

Dipleurosomatidae

SCIENTIFIC NAME	OCCURRENCE	COMMON NAME
Dichotomia cannoides Brooks, 1903	A	
Dipleurosoma collapsum (Mayer, 1910)	A	
Dipleurosoma ochraceum Mayer, 1910	A	
Dipleurosoma typicum Boeck, 1866	A-P	

Melicertidae

SCIENTIFIC NAME	OCCURRENCE	COMMON NAME
+^ *Melicertum octocostatum* (M. Sars, 1835)	A-P-Ac	
Netocertoides brachiatus Mayer, 1900	A	
Orchistomella tentaculata (Mayer, 1910)	A	

Orchistomatidae

SCIENTIFIC NAME	OCCURRENCE	COMMON NAME
Orchistoma pileus (Lesson, 1843)	A	

+Aequoreidae

SCIENTIFIC NAME	OCCURRENCE	COMMON NAME
Aequorea albida A. Agassiz, 1862	A-P	
Aequorea coerulescens (Brandt, 1838)	P	
Aequorea floridana (L. Agassiz, 1862)	A	
+^ *Aequorea forskalea* (Péron and Lesueur, 1810)	A-P	
Aequorea tenuis (A. Agassiz, 1862)	A	
^+ *Aequorea victoria* (Murbach and Shearer, 1902)	P	water jellyfish
Rhacostoma atlanticum L. Agassiz, 1850	A	

Blackfordiidae

SCIENTIFIC NAME	OCCURRENCE	COMMON NAME
Blackfordia manhattensis Mayer, 1910	A-E	
*^ *Blackfordia virginica* Mayer, 1910	A(E,I)-P(E,I)	

Phialellidae

SCIENTIFIC NAME	OCCURRENCE	COMMON NAME
+^ *Phialella fragilis* (Uchida, 1938)	P	
Phialella parvigastra (Mayer, 1900)	P	
+^ *Phialella zappai* Boero, 1987	P	

Occurrence abbreviations: A=Atlantic, P=Pacific, Ac=Arctic, E=Estuarine, F=Freshwater, I=Introduced.
*Change from first edition of 1991. +Not changed, but comment in Part II (Appendix 1).
^Species listed under both their hydroid and medusoid stages.

SCIENTIFIC NAME	OCCURRENCE	COMMON NAME
Laodiceidae		
Laodicea brevigona Allwein, 1967	A	
Laodicea neptuna Mayer, 1900	A	
Laodicea undulata (Forbes and Goodsir, 1853)	A	
Melicertissa mayeri Kramp, 1959	A	
Ptychogena californica Torrey, 1909	P	
^+*Ptychogena lactea* A. Agassiz, 1865	A-P-Ac.	
Staurophora mertensi Brandt, 1838	A-P	whitecross jellyfish
Toxorchis kellneri Mayer, 1910	A	
***Tiarannidae**		
Chromatonema rubrum Fewkes, 1882	A-P	
* *Modeeria rotunda* (Quoy and Gaimard, 1827)	P	
Malagazziidae		
Malagazzia carolinae (Mayer, 1900)	A-E	
Mitrocomidae		
Cyclocanna welshi H. B. Bigelow, 1918	A	
Foersteria purpurea (Foerster, 1923)	P	
Halopsis ocellata (A. Agassiz, 1863)	A	
Mitrocoma cellularia (A. Agassiz, 1865)	P	
Mitrocoma discoidea Torrey, 1909	P	
Mitrocomella cruciata A. Agassiz, 1865	A	
Mitrocomella polydiademata (Romanes, 1876)	A-P-Ac.	
Mitrocomella sinuosa (Foerster, 1923)	P	
Tiaropsidae		
Tiaropsidium kelsei Torrey, 1909	P	
Tiaropsis multicirrata (M. Sars, 1835)	A-P-Ac.	
***Cirrholoveniidae**		
Cirrholovenia tetranema Kramp, 1959	A	
***Eucheilotidae**		
+^*Eucheilota bakeri* (Torrey, 1909)	P	
Eucheilota duodecimalis A. Agassiz, 1862	A	
Eucheilota pardoxica Mayer, 1900	A	
Eucheilota ventricularis McCrady, 1859	A	
***Lovenellidae**		
Lovenella bermudensis (Fewkes, 1883)	A	
^ *Lovenella gracilis* (Clarke, 1882)	A	
***Eirenidae**		
Eirene gibbosa (McCrady, 1859)	A	
Eirene lactea (Mayer, 1900)	A	
Eirene mollis Torrey, 1909)	P	
Eirene pyramidalis (L. Agassiz, 1862)	A	
Eutima coerulea (L. Agassiz, 1862)	A	
Eutima cuculata Brooks, 1883	A	
Eutima gegenbauri (Haeckel, 1864)	A	
Eutima mira McCrady, 1859	A	
Eutima suzannae Allwein, 1967	A	
Eutima variabilis McCrady, 1859	A	
Eutimalphes brownei Torrey, 1909	P	

Occurrence abbreviations: A=Atlantic, P=Pacific, Ac=Arctic, E=Estuarine, F=Freshwater, I=Introduced.
*Change from first edition of 1991. +Not changed, but comment in Part II (Appendix 1).
^Species listed under both their hydroid and medusoid stages.

SCIENTIFIC NAME	OCCURRENCE	COMMON NAME
Eutonina indicans (Romanes, 1876)	P	
Helgicirrha weaveri Allwein, 1967	A	
Phialopsis diegensis Torrey, 1909	P	
***Campanulariidae**		
Clytia discoida (Mayer, 1900)	A	
Clytia folleata (McCrady, 1859)	A	
Clytia gelatinosa (Mayer, 1900)	A	
Clytia globosa (Mayer, 1900)	A	
^ *Clytia gregaria* (L. Agassiz, 1862)	P	
^* *Clytia hemispherica* (Linnaeus, 1767)	A	
^ *Clytia lomae* (Torrey, 1909)	P	
Clytia mccradyi (Brooks, 1888)	A	
^ *Clytia noliformis* auct.	A	
Clytia singularis (Mayer, 1900)	A	
*^ *Obelia dichotoma* (Linnaeus, 1758)	P	
*^ *Obelia longissima* (Pallas, 1766)	P	
^* *Orthopyxis compressa* (Clark, 1877)	P	
+ORDER NARCOMEDUSAE		
Aeginidae		
* *Aegina citrea* Eschscholtz, 1829	A-P	golf tee medusa
Aeginopsis laurentii Brandt, 1838	A-P-Ac	
Aeginura grimaldii Maas, 1904	A-P	
Solmundella bitentaculata (Quoy and Gaimard, 1833)	A-P	
Cuninidae		
Cunina duplicata Maas, 1893	A	
* *Cunina frugifera* Kramp, 1948	P	
* *Cunina globosa* Eschscholtz, 1829	P	
Cunina octonaria McCrady, 1859	A-P	
Cunina peregrina H. B. Bigelow, 1909	A	
Solmissus incisus (Fewkes, 1886)	A-P	
* *Solmissus marshalli* A. Agassiz and Mayer, 1902	P	dinner plate medusa
Polypodiidae		
+ *Polypodium hydriforme* Ussov, 1885	F	
Solmarisidae		
+ *Pegantha clara* H. B. Bigelow, 1909	A	
* *Pegantha laevis* H. B. Bigelow, 1909	P	
* *Pegantha martagon* Haeckel, 1879	A	
Solmaris corona (Keferstein and Ehlers, 1861)	P	
+ORDER TRACHYMEDUSAE		
Geryoniidae		
Geryonia proboscidalis (Forskål, 1775)	A	
Liriope tetraphylla (Chamisso and Eysenhardt, 1821)	A-P	
Halicreatidae		
Botrynema brucei Browne, 1908	A-P	
Botrynema ellinorae (Hartlaub, 1909)	Ac	
Halicreas minimum Fewkes, 1882	A-P	

Occurrence abbreviations: A=Atlantic, P=Pacific, Ac=Arctic, E=Estuarine, F=Freshwater, I=Introduced.
*Change from first edition of 1991. +Not changed, but comment in Part II (Appendix 1).
^Species listed under both their hydroid and medusoid stages.

SCIENTIFIC NAME	OCCURRENCE	COMMON NAME
* *Haliscera bigelowi* Kramp, 1947	A-P	
* *Haliscera conica* Vanhöffen, 1902	P	
* *Haliscera racovitzae* (Maas, 1906)	P	
* *Halitrephes maasi* H. B. Bigelow, 1909	A-P	
* *Halitrephes valdiviae* Vanhöffen, 1912	A	
Ptychogastriidae		
Ptychogastria polaris Allman, 1878	A-P-Ac.	
Rhopalonematidae		
Aglantha digitale (O. F. Müller, 1776)	A-P-Ac	pink helmet
Aglaura hemistoma Péron and Lesueur, 1810	A-P	
* *Benthocodon pedunculata* (H. B. Bigelow, 1913)	P	
* *Colobonema sericeum* Vanhöffen, 1902	A-P	silky medusa
Colobonema typicum (Maas, 1897)	P	
Crossota alba H. B. Bigelow, 1913	P	
Crossota rufobrunnea (Kramp, 1913)	A-P	
Homoeonema platygonon Browne, 1903	Ac.	
Pantachogon haeckeli Maas, 1893	A-P-Ac.	
Persa incolorata McCrady, 1859	A	
Rhopalonema funerarium Vanhöffen, 1902	A	
Rhopalonema velatum Gegenbaur, 1856	A-P	
Sminthea arctica Hartlaub, 1909	Ac.	
* *Tetrorchis erythrogaster* Bigelow, 1909	P	
* *Vampyrocrossota childressi* Thuesen, 1993	P	

*Subclass Siphonophorae—Siphonophores

ORDER CYSTONECTAE

SCIENTIFIC NAME	OCCURRENCE	COMMON NAME
Physaliidae		
Physalia physalis (Linnaeus, 1758)	A-P	Portuguese man o' war
+Rhizophysidae		
* *Bathyphysa conifera* (Studer, 1878)	A	
Rhizophysa eysenhardti Gegenbaur, 1859	P	
Rhizophysa filiformis (Forskål, 1775)	A	

ORDER PHYSONECTAE

SCIENTIFIC NAME	OCCURRENCE	COMMON NAME
Apolemiidae		
+ *Apolemia uvaria* (Leseuer, 1815)	A-P	
Ramosia vitiazi Stepanjants, 1967	P	
+Agalmatidae		
* *Agalma elegans* (M. Sars, 1846)	A-P	
Agalma okeni Eschscholtz, 1825	A-P	
Cordagalma cordiforme Totton, 1932	A-P	
Erenna richardi Bedot, 1904	A-P	
* *Frillagalma vityazi* Daniel, 1966	A-P	
* *Halistemma amphytridis* (Lesueur and Petit, 1807)	A-P	
Halistemma cupulifera Lens and van Riemsdijk, 1908	A	
Halistemma rubrum (Vogt, 1852)	A-P	
* *Halistemma striata* Totton, 1965	A	
* *Halistemma transliratum* Pugh and Youngbluth, 1988	A	

Occurrence abbreviations: A=Atlantic, P=Pacific, Ac=Arctic, E=Estuarine, F=Freshwater, I=Introduced.
*Change from first edition of 1991. +Not changed, but comment in Part II (Appendix 1).
^Species listed under both their hydroid and medusoid stages.

SCIENTIFIC NAME	OCCURRENCE	COMMON NAME
Lychnagalma utricularia (Claus, 1879)	A	
Marrus orthocanna (Kramp, 1942)	A-P	
Nanomia bijuga (delle Chiaje, 1841)	A-P	
+ *Nanomia cara* A. Agassiz, 1865	A-P	
***Pyrostephidae**		
* *Bargmannia amoena* Pugh, 1999	A	
+ *Bargmannia elongata* Totton, 1954	A-P	
* *Bargmannia lata* Mapstone, 1998	P	
Physophoridae		
Physophora hydrostatica Forskål, 1775	A-P	
Athorybiidae		
Athorybia lucida Biggs, 1978	A	
Athorybia rosacea (Forskål, 1775)	A-P	
Rhodaliidae		
Angelopsis globosa Fewkes, 1886	A	
Dromalia alexandri H.B. Bigelow, 1911	P	
Forskaliidae		
Forskalia edwardsi Kölliker, 1853	A-P	
Forskalia tholoides Haeckel, 1888	A	
ORDER CALYCOPHORAE		
Prayidae		
Amphicaryon acaule Chun, 1888	A-P	
Amphicaryon ernesti Totton, 1954	A-P	
* *Craseoa lathetica* Pugh and Harbison, 1987	A	
* *Desmophyes annectens* Haeckel, 1888	A-P	
* *Desmophyes haematogaster* Pugh, 1992	A	
Lilyopsis rosea Chun, 1885	P	
* *Maresearsia praeclara* Totton, 1954	P	
* *Mistoprayina fragosa* Pugh and Harbison, 1987	A	
* *Nectadamas diomedeae* (H. B. Bigelow, 1911)	A-P	
Nectopyramis natans (H. B. Bigelow, 1911)	P	
Nectopyramis thetis H. B. Bigelow, 1911	P	
Praya dubia (Quoy and Gaimard, 1827)	A-P	
* *Praya reticulata* (H. B. Bigelow, 1911)	A-P	
* *Prayola urinatrix* Pugh and Harbison, 1987	A	
Rosacea cymbiformis (delle Chiaje, 1822)	A-P	
Rosacea flaccida Biggs, Pugh and Carré, 1978	A	
* *Rosacea limbata* Pugh and Youngbluth, 1988	A	
+ *Rosacea plicata* sensu H. B. Bigelow, 1911	P	
* *Rosacea repanda* Pugh and Youngbluth, 1988	A	
Stephanophyes superba Chun, 1888	A-P	
Hippopodiidae		
Hippopodius hippopus (Forskål, 1776)	A-P	
* *Vogtia glabra* H. B. Bigelow, 1918	A	
Vogtia pentacantha Kölliker, 1853	A-P	
+ *Vogtia serrata* (Moser, 1925)	A-P	
Vogtia spinosa Keferstein and Ehlers, 1861	A-P	

Occurrence abbreviations: A=Atlantic, P=Pacific, Ac=Arctic, E=Estuarine, F=Freshwater, I=Introduced.
*Change from first edition of 1991. +Not changed, but comment in Part II (Appendix 1).
^Species listed under both their hydroid and medusoid stages.

SCIENTIFIC NAME	OCCURRENCE	COMMON NAME
Diphyidae		
Chelophyes appendiculata (Eschscholtz, 1829)	A-P	
Chelophyes contorta (Lens and van Riemsdijk, 1908)	P	
Dimophyes arctica (Chun, 1897)	A-P	
Diphyes bojani (Eschscholtz, 1829)	A-P	
Diphyes dispar Chamisso and Eysenhardt, 1821	A-P	
Eudoxoides mitra (Huxley, 1859)	A-P	
Eudoxoides spiralis (H.B. Bigelow, 1911)	A-P	
* *Gilia reticulata* (Totton, 1954)	P	
Lensia achilles Totton, 1941	A-P	
Lensia ajax Totton, 1941	P	
+ *Lensia baryi* Totton 1965	P	
Lensia campanella (Moser, 1925)	A-P	
Lensia challengeri Totton, 1954	P	
Lensia conoidea (Keferstein and Ehlers, 1860)	A-P	
* *Lensia cossack* Totton, 1941	A-P	
* *Lensia exeter* Totton, 1941	P	
Lensia fowleri (H.B. Bigelow, 1911)	A-P	
Lensia grimaldi (Leloup, 1933)	P	
Lensia havock Totton, 1941	P	
Lensia hostile Totton, 1941	P	
Lensia hotspur Totton, 1941	A-P	
Lensia lelouveteau Totton, 1941	P	
Lensia meteori (Leloup, 1934)	A-P	
Lensia multicristata (Moser, 1925)	A-P	
* *Lensia subtilis* (Chun, 1886)	A-P	
* *Lensia subtiloides* (Lens and van Riemsdijk, 1908)	P	
Muggiaea atlantica Cunningham, 1892	A-P	
Muggiaea bargmannae Totton, 1954	A-P	
Muggiaea kochi (Will, 1854)	A	
Sulculeolaria biloba (M. Sars, 1846)	A-P	
Sulculeolaria chuni (Lens and van Riemsdijk, 1908)	A-P	
Sulculeolaria monoica (Chun, 1888)	A-P	
Sulculeolaria quadrivalvis Blainville, 1834	A-P	
* *Sulculeolaria turgida* (Gegenbaur, 1853)	A-P	
Clausophyidae		
* *Chuniphyes moserae* Totton, 1954	A-P	
Chuniphyes multidentata Lens and van Riemsdijk, 1908	A-P	
Clausophyes galeata Lens and van Riemsdijk, 1908	P	
* *Clausophyes moserae* Margulis, 1988	A-P	
* *Clausophyes ovata* (Keferstein and Ehlers, 1860)	A-P	
+ *Crystallophyes amygdalina* Moser, 1925	P	
Heteropyramis maculata Moser, 1925	P	
Sphaeronectidae		
Sphaeronectes gracilis (Claus, 1873)	A-P	
* *Sphaeronectes irregularis* (Claus, 1873)	P	
Abylidae		
+ *Abyla bicarinata* Moser, 1925	P	
* *Abyla haeckeli* Lens and van Riemsdijk, 1908	A-P	
Abyla trigona Quoy and Gaimard, 1827	A	

Occurrence abbreviations: A=Atlantic, P=Pacific, Ac=Arctic, E=Estuarine, F=Freshwater, I=Introduced.
*Change from first edition of 1991. +Not changed, but comment in Part II (Appendix 1).
^Species listed under both their hydroid and medusoid stages.

SCIENTIFIC NAME	OCCURRENCE	COMMON NAME
Abylopsis eschscholtzi (Huxley, 1859)	A-P	
Abylopsis tetragona (Otto, 1823)	A-P	
Bassia bassensis (Quoy and Gaimard, 1833)	A-P	
Ceratocymba dentata (H.B. Bigelow, 1918)	P	
+ *Ceratocymba leuckarti* (Huxley, 1859)	A	
Ceratocymba sagittata (Quoy and Gaimard, 1827)	A	
+ *Enneagonum hyalinum* Quoy and Gaimard, 1827	A	

Class Anthozoa

SUBCLASS CERIANTIPATHARIA

ORDER ANTIPATHARIA—BLACK CORALS, THORNY CORALS

*Antipathidae

SCIENTIFIC NAME	OCCURRENCE	COMMON NAME
Antipathes americana Duchassaing and Michelotti, 1860	A	
* *Antipathes atlantica* Gray, 1857	A	grey sea fan black coral
* *Antipathes barbadensis* Brook, 1889	A	scraggly bottle brush black coral
* *Antipathes caribbeana* Opresko, 1996	A	Kings coral
* *Antipathes expansa* Opresko and Cairns, 1992	A	
Antipathes furcata Gray, 1857	A	
* *Antipathes gracilis* Gray, 1860	A	orange sea fan black coral
* *Antipathes hirta* Gray, 1857	A	bottle brush bush black coral
* *Antipathes lenta* Pourtalès, 1871	A	hair net black coral
Antipathes pedata Gray, 1857	A	
* *Antipathes pennacea* Pallas, 1766	A	feather black coral
* *Antipathes tanacetum* Pourtalès, 1880	A	bottle brush black coral
* *Antipathes thamnea* Warner, 1981	A	
Aphanipathes abietina (Pourtalès, 1874)	A	
Aphanipathes humilis (Pourtalès, 1867)	A	
Aphanipathes thyroides (Pourtalès, 1880)	A	
Parantipathes tetrasticha (Pourtalès, 1868)	A	
Stichopathes lutkeni (Brook, 1889)	A	black wire coral

*Leiopathidae

SCIENTIFIC NAME	OCCURRENCE	COMMON NAME
* *Leiopathes glaberrima* (Esper, 1788)	A	smooth black coral

ORDER CERIANTHARIA—TUBE-DWELLING ANEMONES

Cerianthidae

SCIENTIFIC NAME	OCCURRENCE	COMMON NAME
* *Ceriantheomorphe brasiliensis* Carlgren, 1931	A	
+ *Ceriantheopsis americana* (Verrill, 1862)	A	
Cerianthus borealis Verrill, 1873	A	
* *Isodactylactis borealis* Widersten, 1998	A	
Pachycerianthus aestuarii (Torrey and Kleeburger, 1909)	P	
+ *Pachycerianthus fimbriatus* McMurrich, 1910	P	
Pachycerianthus johnsoni (Torrey and Kleeburger, 1909)	P	
* *Synarachnactis brachiolata* (A. Agassiz, 1862)	A-Ac	

Botrucnidiferidae

SCIENTIFIC NAME	OCCURRENCE	COMMON NAME
Botruanthus benedeni (Torrey and Kleeburger, 1909)	P	

Occurrence abbreviations: A=Atlantic, P=Pacific, Ac=Arctic, E=Estuarine, F=Freshwater, I=Introduced.
*Change from first edition of 1991. +Not changed, but comment in Part II (Appendix 1).
^Species listed under both their hydroid and medusoid stages.

SCIENTIFIC NAME	OCCURRENCE	COMMON NAME
Subclass Alcyonaria = Octocoralia		
ORDER ALCYONACEA—SOFT CORALS		
Clavulariidae		
Carijoa operculata (Bayer, 1961)	A	
* *Carijoa riisei* (Duchassaing and Michelotti, 1860)	A-P	white telesto
Clavularia modesta (Verrill, 1874)	A	
* *Cryptophyton goddarti* Williams, 2000	P	
Telesto flavula Deichmann, 1936	A	
* *Telesto fruticulosa* Dana, 1846	A	orange telesto
Telesto nelleae Bayer, 1961	A	
Telesto sanguinea Deichmann, 1936	A	
SUBORDER ALCYONIINA—TRUE SOFT CORALS		
Alcyoniidae		
Alcyonium digitatum Linnaeus, 1758	A	
* *Alcyonium rudyi* Verseveldt and van Ofwegen, 1992	P	
Alcyonium sidereum Verrill, 1922	A	
Drifa glomerata (Verrill, 1869)	A	
Duva florida (Rathke, 1806)	A	
* *Gersemia rubiformis* (Ehrenberg, 1834)	A-Ac	sea strawberry
Pseudodrifa nigra (Pourtalès, 1868)	A	
Nidaliidae		
Nidalia occidentalis Gray, 1835	A	
* *Siphonogorgia agassizii* (Deichmann, 1936)	A	
SUBORDER SCLERAXONIA		
***Keroeididae**		
* *Thelogorgia studeri* Bayer, 1991	A	
Briaridae		
* *Briareum asbestinum* (Pallas, 1766)	A	corky seafinger or deadman's fingers
Anthothelidae		
Anthopodium rubens Verrill, 1872	A	
Anthothela grandiflora (Sars, 1856)	A	
Anthothela tropicalis Bayer, 1961	A	
* *Diodogorgia nodulifera* (Hargitt and Rogers, 1901)	A	colorful sea rod
* *Erythropodium caribaeorum* (Duchassaing and Michelotti, 1860)	A	encrusting gorgonian
Iciligorgia schrammi Duchassaing, 1870	A	deep-water sea fan
* *Titanideum frauenfeldii* (Kölliker, 1865)	A	brilliant sea fingers
Paragorgiidae		
* *Paragorgia arborea* (Linnaeus, 1758)	A-P	bubble gum coral
SUBORDER HOLAXONIA		
Plexauridae		
* *Bebryce grandis* Deichmann, 1936	A	
* *Bebryce parastellata* Deichmann, 1936	A	
Eunicea asperula Milne Edwards and Haime, 1857	A	
* *Eunicea calyculata* (Ellis and Solander, 1786)	A	warty sea rod
* *Eunicea fusca* Duchassaing and Michelotti, 1860	A	doughnut sea rod

SCIENTIFIC NAME	OCCURRENCE	COMMON NAME
Eunicea knighti Bayer, 1961	A	
Eunicea laciniata Duchassaing and Michelotti, 1860	A	
* *Eunicea mammosa* Lamouroux, 1816	A	swollen-knob candelabrum
Eunicea palmeri Bayer, 1961	A	
* *Eunicea succinea* (Pallas, 1766)	A	shelf-knob sea rod
Eunicea tourneforti Milne Edwards and Haime, 1857	A	
* *Hypnogorgia pendula* Duchassaing and Michelotti, 1864	A	
* *Lytreia plana* (Deichmann, 1936)	A	
Muricea atlantica (Riess, 1919 in Kükenthal)	A	
Muricea californica Aurivillius, 1931	P	
* *Muricea elongata* Lamouroux, 1821	A	orange spiny sea rod
Muricea fruticosa Verrill, 1869	P	
* *Muricea laxa* Verrill, 1864	A	delicate spiny sea rod
* *Muricea muricata* (Pallas, 1766)	A	spiny sea fan
* *Muricea pendula* Verrill, 1864	A	pinnate spiny sea fan
* *Muriceides hirtus* (Pourtalès, 1868)	A	
* *Muriceopsis flavida* (Lamarck, 1815)	A	rough sea plume
Muriceopsis petila Bayer, 1961	A	
* *Paramuricea grandis* Verrill, 1883	A	
* *Paramuricea placomus* (Linnaeus, 1758)	A	
* *Placogorgia mirabilis* Deichmann, 1936	A	
* *Placogorgia tenuis* (Verrill, 1883)	A	
* *Plexaura flexuosa* Lamouroux, 1821	A	bent sea rod
Plexaura homomalla (Esper, 1792)	A	black sea rod
* *Plexaurella dichotoma* (Esper, 1791)	A	slit-pore sea rod
Plexaurella fusifera Kunze, 1916	A	
Plexaurella grisea Kunze, 1916	A	
* *Plexaurella nutans* (Duchassaing and Michelotti, 1860)	A	giant slit-pore sea rod
Psammogorgia teres Verrill, 1868	P	
* *Pseudoplexaura flagellosa* (Houttuyn, 1772)	A	porous sea rod
Pseudoplexaura porosa (Houttuyn, 1772)	A	
Pseudoplexaura wagenaari (Stiasny, 1941)	A	
* *Scleracis guadalupensis* (Duchassaing and Michelotti, 1860)	A	
* *Scleracis pumila* Reiss, 1919	A	
* *Spinimuricea atlantica* (Johnson, 1862)	A	
* *Swiftia casta* (Verrill, 1883)	A	
* *Swiftia exserta* Duchassaing and Michelotti, 1864	A	red polyp octocoral
* *Swiftia kofoidi* (Nutting, 1909)	P	
* *Swiftia spauldingi* (Nutting, 1909)	P	
Thesea citrina Deichmann, 1936	A	
Thesea grandiflora Deichmann, 1936	A	
Thesea guadalupensis (Duchassaing and Michelotti, 1860)	A	
* *Thesea nivea* Deichmann, 1936	A	white eye sea spray
Thesea rugosa Deichmann, 1936	A	
Thesea solitaria (Pourtalès, 1868)	A	
* *Villogorgia nigrescens* Duchassaing and Michelotti, 1860	A	
Gorgoniidae		
Adelogorgia phyllosclera Bayer, 1958	P	
Eugorgia ampla (Verrill, 1864)	P	
Eugorgia forreri Studer, 1883	P	
Eugorgia nobilis Verrill, 1868	P	

Occurrence abbreviations: A=Atlantic, P=Pacific, Ac=Arctic, E=Estuarine, F=Freshwater, I=Introduced.
*Change from first edition of 1991. +Not changed, but comment in Part II (Appendix 1).
^Species listed under both their hydroid and medusoid stages.

SCIENTIFIC NAME	OCCURRENCE	COMMON NAME
Eugorgia rubens Verrill, 1868	P	
Gorgonia flabellum Linnaeus, 1758	A	Venus sea fan
Gorgonia ventalina Linnaeus, 1758	A	common sea fan
Leptogorgia californica Verrill, 1868	P	
* *Leptogorgia cardinalis* (Bayer, 1961)	A	
* *Leptogorgia caryi* Verrill, 1868	P	
* *Leptogorgia chilensis* Verrill, 1868	P	
Leptogorgia euryale (Bayer, 1952)	A	
* *Leptogorgia hebes* Verrill, 1869	A	regal sea fan
Leptogorgia medusa (Bayer, 1952)	A	
* *Leptogorgia miniata* (Milne-Edwards and Haime, 1857)	A	carmine sea spray
* *Leptogorgia punicea* (Milne Edwards and Haime, 1857)	A	
Leptogorgia setacea (Pallas, 1766)	A	
Leptogorgia stheno (Bayer, 1952)	A	
* *Leptogorgia virgulata* (Lamarck, 1815)	A	colorful sea whip
Pacifigorgia pulchra (Verrill, 1870)	P	
Pseudopterogorgia acerosa (Pallas, 1766)	A	purple sea plume
Pseudopterogorgia americana (Gmelin, 1791)	A	slimy sea plume
* *Pseudopterogorgia bipinnata* (Verrill, 1864)	A	bipinnate sea plume
* *Pseudopterogorgia elisabethae* Bayer, 1961	A	sea plume
Pseudopterogorgia kallos Bielchowsky, 1918	A	
Pseudopterogorgia rigida (Bielchowsky, 1929)	A	
Pterogorgia anceps (Pallas, 1766)	A	angular sea whip
Pterogorgia citrina (Esper, 1792)	A	yellow sea whip
* *Pterogorgia guadalupensis* Duchassaing and Michelin, 1846	A	grooved-blade sea whip

*Suborder Calcaxonia

Ellisellidae

SCIENTIFIC NAME	OCCURRENCE	COMMON NAME
Ellisella barbadensis (Duchassaing and Michelotti, 1864)	A	devil's sea whip
Ellisella elongata (Pallas, 1766)	A	long sea whip
Ellisella grandis (Verrill, 1901)	A	
* *Ellisella schmitti* Bayer, 1961	A	bushy sea whip
* *Nicella goreaui* Bayer, 1973	A	orange deep water fan
Nicella guadalupensis (Duchassaing and Michelotti, 1860)	A	
* *Riisea paniculata* Duchassaing and Michelotti, 1860	A	

Primnoidae

SCIENTIFIC NAME	OCCURRENCE	COMMON NAME
* *Arthrogorgia utinomii* Bayer, 1996	P	
Callogorgia kinoshitae Kükenthal, 1913	P	
Callogorgia verticillata (Pallas, 1766)	A	
* *Fanellia compressa* (Verrill, 1865)	P	
* *Fanellia fraseri* Hickson, 1915	P	
Plumarella pourtalesii (Verrill, 1883)	A	
* *Primnoa resedaeformis* (Gunnerus, 1763)	A	red trees

Family Isididae

SCIENTIFIC NAME	OCCURRENCE	COMMON NAME
Stenisis humilis (Deichmann, 1936)	A	

SCIENTIFIC NAME	OCCURRENCE	COMMON NAME
ORDER PENNATULACEA—SEA PENS		
SUBORDER SESSILIFLORAE		
Renillidae		
Renilla koellikeri Pfeffer, 1886	P	
Renilla muelleri Kölliker, 1872	A	
Renilla reniformis (Pallas, 1766)	A	sea pansy
Kophobelemnidae		
Kophobelemnon stelliferum (Müller, 1776)	A-P	
Anthoptilidae		
Anthoptilum grandiflorum (Verrill, 1879)	A-P	
Family Funiculinidae		
Funiculina quadrangularis (Pallas, 1766)	A-P	
Family Protoptilidae		
Protoptilum thomsoni Kölliker, 1872	A	
Family Stachyptilidae		
Stachyptilum superbum Studer, 1894	A	
SUBORDER SUBSELLIFLORAE		
***Virgulariidae**		
* *Acanthoptilum agassizi* Bayer (1957)	A	
Acanthoptilum album Nutting, 1909	P	
Acanthoptilum annulatum Nutting, 1909	P	
Acanthoptilum gracile (Gabb, 1862)	P	
* *Acanthoptilum oligacis* Bayer, 1957	A	
Acanthoptilum scalpellifolium Moroff, 1902	P	
* *Halipteris christii* (Koren and Danielssen, 1847)	A	
* *Halipteris finmarchica* (Sars, 1851)	A	
* *Stylatula antillarum* Kölliker, 1872	A	
Stylatula elegans (Danielssen, 1860)	A	
* *Stylatula elongata* (Gabb, 1862)	P	slender sea pen
Stylatula gracilis Verrill, 1864	P	
Virgularia agassizii Studer, 1894	P	
Virgularia mirabilis (Müller, 1776)	A	
Virgularia presbytes Bayer, 1955	A	
Virgularia reinwardti Herklots, 1858	P	
Pennatulidae		
Pennatula aculeata Danielssen, 1860	A	
Pennatula borealis M. Sars, 1846	A	
Pennatula phosphorea Linnaeus, 1758	A-P	
* *Ptilosarcus gurneyi* (Gray, 1860)	P	
* *Ptilosarcus undulatus* (Verrill, 1865)	P	

Subclass Zoantharia (=Hexacorallia)

ORDER ACTINIARIA—ANEMONES

SCIENTIFIC NAME	OCCURRENCE	COMMON NAME
Actiniidae		
Anemonia sargassensis Hargitt, 1908	A	sargassum anemone
Anthopleura artemisia (Pickering in Dana, 1848)	P	
Anthopleura elegantissima (Brandt, 1835)	P	clonal anemone

Occurrence abbreviations: A=Atlantic, P=Pacific, Ac=Arctic, E=Estuarine, F=Freshwater, I=Introduced.
*Change from first edition of 1991. +Not changed, but comment in Part II (Appendix 1).
^Species listed under both their hydroid and medusoid stages.

SCIENTIFIC NAME	OCCURRENCE	COMMON NAME
Anthopleura krebsi Duchassaing and Michelotti, 1860	A	rock anemone
* *Anthopleura sola* Pearce and Francis, 2000	P	sun anemone
Anthopleura varioarmata Watzl, 1922	A	
Anthopleura xanthogrammica (Brandt, 1835)	P	giant green anemone
Aulactinia incubans Dunn, Chia, and Levine, 1980	P	incubating anemone
Bolocera tuediae (Johnston, 1832)	A	
Bunodactis capitata (Verrill, 1864)	A	
Bunodactis stella (Verrill, 1864)	A	
Bunodactis texaensis Carlgren and Hedgpeth, 1952	A	
Bunodosoma cavernatum (Bosc, 1802)	A	
* *Bunodosoma granuliferum* (Lesueur, 1817)	A	
Condylactis gigantea (Weinland, 1860)	A	giant Caribbean anemone
Cribrinopsis fernaldi Siebert and Spaulding, 1976	P	chevron-tentacle anemone
Cribrinopsis williamsi Carlgren, 1940	P	Williams anemone
Epiactis arctica (Verrill, 1868)	Ac	Arctic brooding anemone
Epiactis fernaldi Fautin and Chia, 1986	P	Fernald brooding anemone
Epiactis lisbethae Fautin and Chia, 1986	P	giant brooding anemone
Epiactis prolifera Verrill, 1869	P	brooding anemone
Epiactis ritteri Torrey, 1902	P	sandy anemone
Leipsiceras pollens (McMurrich, 1898)	A	
Urticina columbiana Verrill, 1922	P	crusty red anemone
Urticina coriacea (Cuvier, 1798)	P	leathery anemone
Urticina crassicornis (Müller, 1776)	P-Ac	mottled anemone
Urticina felina (Linnaeus, 1767)	A	
Urticina lofotensis (Danielssen, 1890)	P	spotted red anemone
* *Urticina mcpeaki* Hauswaldt and Pearson, 1999	P	McPeak anemone
* *Urticina piscivora* (Sebens and Laakso, 1978)	P	velvety red anemone
Actinoscyphiidae		
* *Actinoscyphia saginata* (Verrill, 1882)	A	Venus fly-trap anemone
Actinostolidae		
Actinostola callosa (Verrill, 1882)	A	
Antholoba perdix (Verrill, 1882)	A	
Paranthus rapiformis (Lesueur, 1817)	A	
* *Sicyonis obesa* (Carlgren, 1934)	A	
Stomphia coccinea (Müller, 1776)	A-P	swimming anemone
Stomphia didemon Siebert, 1973	P	cowardly anemone
* *Stomphia pacifica* Ross and Zamponi, 1992	P	
Aiptasiomorphidae		
Aiptasiomorpha texaensis Carlgren and Hedgpeth, 1952	A	
Aiptasiidae		
* *Aiptasia californica* Carlgren, 1952	P	
Aiptasia eruptaurantia (Field, 1949)	A	
Aiptasia pallida (Verrill, 1864)	A	pale anemone
Aiptasia tagetes (Duchassaing and Michelotti, 1864)	A	
* *Bartholomea annulata* (Lesueur, 1817)	A	corkscrew anemone
Aliciidae		
* *Alicia mirabilis* Johnson, 1861	A	berried anemone
* *Lebrunia coralligens* (Wilson, 1890)	A	hidden anemone
* *Lebrunia danae* (Duchassaing and Michelotti, 1860)	A	branching anemone

Occurrence abbreviations: A=Atlantic, P=Pacific, Ac=Arctic, E=Estuarine, F=Freshwater, I=Introduced.
*Change from first edition of 1991. +Not changed, but comment in Part II (Appendix 1).
^Species listed under both their hydroid and medusoid stages.

SCIENTIFIC NAME	OCCURRENCE	COMMON NAME
Andwakiidae		
Andwakia isabellae Carlgren and Hedgpeth, 1952	A	
Boloceroididae		
Viatrix eugenia (Duchassaing and Michelotti, 1864)	A	
* *Viatrix globulifera* (Duchassaing, 1850)	A	turtle-grass anemone
Viatrix reclinata (Bosc, 1802)	A	
Condylanthidae		
Charisea saxicola Torrey, 1902	P	Alaskan anemone
Diadumenidae		
Diadumene cincta Stephenson, 1925	P	orange anemone
* *Diadumene franciscana* Hand, 1956	P	San Francisco anemone
Diadumene leucolena (Verrill, 1866)	A-P	white anemone
* *Diadumene lighti* Hand, 1956	P	Lights anemone
* *Diadumene lineata* (Verrill, 1870)	P	orangestriped green anemone
Edwardsiidae		
* *Drillactis pallida* (Agassiz in Verrill, 1864)	A	
Edwardsia californica (McMurrich, 1913)	P	Californian burrowing anemone
* *Edwardsia elegans* Verrill, 1869	A	elegant burrowing anemone
* *Edwardsia leidyi* Verrill, 1898	A	Leidys burrowing anemone
Edwardsia sipunculoides (Stimpson, 1853)	A-P	sipunculid-like anemone
* *Edwardsia sulcata* Verrill, 1864	A	
Fagesia lineata (Verrill, 1874)	A	
* *Nematostella vectensis* Stephenson, 1935	A-P	starlet anemone
Halcampidae		
* *Cactosoma arenaria* Carlgren, 1931	P	prickly anemone
Halcampa crypta Siebert and Hand, 1974	P	cryptic burrowing anemone
* *Halcampa decemtentaculata* Hand, 1955	P	ten-tentacled burrowing anemone
Halcampa duodecimcirrata (Sars, 1851)	A	twelve-tentacled burrowing anemone
Halcampoididae		
* *Halcampoides purpureus* (Studer, 1879)	P	
* *Pentactinia californica* Carlgren, 1900	P	
Halcuriidae		
Halcurias pilatus McMurrich, 1893	A	
Haloclavidae		
* *Anemonactis mazeli* (Jourdan, 1880)	P	
Bicidium aequoreae McMurrich, 1913	P	parasitic anemone
Haloclava producta (Stimpson, 1856)	A	
Harenactis attenuata Torrey, 1902	P	giant burrowing anemone
Ilyanthus chloropsis Verrill, 1864	A	
Ilyanthus laevis Verrill, 1864	A	
* *Peachia parasitica* (Agassiz, 1859)	A	parasitic anemone
Peachia quinquecapitata McMurrich, 1913	P	twelve-tentacled parasitic anemone
Hormathiidae		
Actinauge longicornis (Verrill, 1882)	A	
Actinauge verrillii McMurrich, 1893	A-P-Ac	reticulate anemone
Adamsia sociabilis Verrill, 1882	A	
Amphianthus californicus Carlgren, 1936	P	

Occurrence abbreviations: A=Atlantic, P=Pacific, Ac=Arctic, E=Estuarine, F=Freshwater, I=Introduced.
*Change from first edition of 1991. +Not changed, but comment in Part II (Appendix 1).
^Species listed under both their hydroid and medusoid stages.

SCIENTIFIC NAME	OCCURRENCE	COMMON NAME
Amphianthus nitidus (Verrill, 1899)	A	
Calliactis tricolor (Lesueur, 1817)	A	
Chondrophellia coronata (Verrill, 1883)	A	
* *Hormathia digitata* (Müller, 1776)	P	
Hormathia nodosa (Fabricius, 1780)	A-Ac	rugose anemone
Phelliactis americana Widersten, 1976	A	
Stephanauge acanellae (Verrill, 1883)	A	
Stephanauge annularis Carlgren, 1936	P	signet ring anemone
Stephanauge nexilis (Verrill, 1883)	A	
Stephanauge spongicola (Verrill, 1883)	A	
Isanthidae		
Zaolutus actius Hand, 1955	P	wormy anemone
Isophelliidae		
Flosmaris grandis Hand and Bushnell, 1967	P	white burrowing anemone
Liponematidae		
* *Liponema brevicornis* (McMurrich, 1893)	P	tentacle-shedding anemone
* *Liponema multicornis* (Verrill, 1879)	A	
Metridiidae		
* *Metridium exilis* Hand, 1956	P	pink clonal anemone
* *Metridium farcimen* (Telesius, 1809)	P	gigantic anemone
* *Metridium senile* (Linnaeus, 1761)	A-P	cloned plumose anemone
Minyadidae		
Minyas olivacea (Lesueur, 1817)	A	
***Nemanthidae**		
* *Nemanthus californicus* Carlgren, 1940	P	
Phymanthidae		
* *Epicystis crucifera* (Lesueur, 1817)	A	beaded anemone
Sagartiidae		
* *Actinothoe californica* Carlgren, 1940	P	
Actinothoe gracillima (McMurrich, 1887)	A	
* *Actinothoe modesta* (Verrill, 1866)	A	
* *Actinothoe pustulata* (McMurrich, 1887)	A	
Botryon tuberculatus Carlgren and Hedgpeth, 1952	A	
* *Sagartia catalienensis* McPeak, 1968	P	Catalina anemone
* *Sagartiogeton verrilli* Carlgren, 1942	A	
Stichodactylidae		
* *Stichodactyla helianthus* (Ellis, 1768)	A	Caribbean sun anemone

ORDER ZOANTHIDEA—ZOANTHIDS

SCIENTIFIC NAME	OCCURRENCE	COMMON NAME
Zoanthidae		
Palythoa mammillosa (Ellis and Solander, 1786)	A	knobby zoanthidean
Palythoa nigricans McMurrich, 1898	A	
Palythoa texaensis Carlgren and Hedgepth, 1952	A	
* *Zoanthus pulchellus* Duchassaing and Michelotti, 1860	A	mat zoanthid

Occurrence abbreviations: A=Atlantic, P=Pacific, Ac=Arctic, E=Estuarine, F=Freshwater, I=Introduced.
*Change from first edition of 1991. +Not changed, but comment in Part II (Appendix 1).
^Species listed under both their hydroid and medusoid stages.

SCIENTIFIC NAME	OCCURRENCE	COMMON NAME
Epizoanthidae		
Epizoanthus abyssorum Verrill, 1885	A	
* *Epizoanthus americanus* Verrill, 1864	A	American zoanthis
Epizoanthus hians McMurrich, 1898	A	
Epizoanthus incrustatus (Duben and Koren, 1847)	A	
Epizoanthus induratum Cutress and Pequegnat, 1960	P	luminescent zoanthid
Epizoanthus leptoderma Cutress and Pequegnat, 1960	P	elongate zoanthid
* *Epizoanthus paguriphilus* Verrill, 1882	A	hermit crab zoanthid
Epizoanthus scotinus Wood, 1958	P	orange zoanthid
Parazoanthidae		
Parazoanthus lucificum Cutress and Pequegnat, 1960	P	luminescent parazoanthid
ORDER CORALLIMORPHARIA		
Actinodiscidae		
* *Discosoma carlgreni* (Watzl, 1922)	A	forked-tentacle corallimorpharian
* *Discosoma neglectum* (Duchassaing and Michelotti, 1860)	A	umbrella corallimorpharian
* *Discosoma sanctithomae* (Duchassaing and Michelotti, 1860)	A	warty corallimorpharian
Corallimorphidae		
Corynactis californica Carlgren, 1936	P	strawberry corallimorpharian
Corynactis delawarei Widersten, 1976	A	
Corynactis parvula Duchassaing and Michelotti, 1860	A	
* *Ricordea florida* Duchassaing and Michelotti, 1860	A	Florida corallimorpharian
ORDER SCLERACTINIA-STONY CORALS		
SUBORDER ASTROCOENIINA		
Astrocoeniidae		
Stephanocoenia michelinii Milne Edwards and Haime, 1848	A	blushing star coral
Pocilloporidae		
Madracis asperula Milne Edwards and Haime, 1849	A	
Madracis brueggemanni (Ridley, 1881)	A	
Madracis decactis (Lyman, 1859)	A	ten-ray star coral
Madracis formosa Wells, 1973	A	eight-ray finger coral
+ *Madracis mirabilis* sensu Wells, 1973	A	yellow pencil coral
Madracis myriaster (Milne Edwards and Haime, 1849)	A	striate finger coral
* *Madracis pharensis* forma *luciphila* Wells, 1973	A	encrusting star coral
* *Madracis pharensis* forma *pharensis* (Heller, 1868)	A	cave star coral
Acroporidae		
Acropora cervicornis (Lamarck, 1816)	A	staghorn coral
Acropora palmata (Lamarck, 1816)	A	elkhorn coral
Acropora prolifera (Lamarck, 1816)	A	fused staghorn coral
SUBORDER FUNGIINA		
Agariciidae		
Agaricia agaricites agaricites (Linnaeus, 1758)	A	lettuce coral
Agaricia agaricites carinata Wells, 1973	A	keeled lettuce coral
Agaricia agaricites danai Milne Edwards and Haime, 1860	A	scaled lettuce coral
Agaricia agaricites purpurea (Lesueur, 1821)	A	purple lettuce coral
Agaricia fragilis contracta Wells, 1973	A	constricted leaf coral

Occurrence abbreviations: A=Atlantic, P=Pacific, Ac=Arctic, E=Estuarine, F=Freshwater, I=Introduced.
*Change from first edition of 1991. +Not changed, but comment in Part II (Appendix 1).
^Species listed under both their hydroid and medusoid stages.

SCIENTIFIC NAME	OCCURRENCE	COMMON NAME
Agaricia fragilis fragilis Dana, 1846	A	fragile saucer coral
Agaricia humilis Verrill, 1901	A	low-relief lettuce coral
Agaricia lamarcki Milne Edwards and Haime, 1851	A	Lamarck's sheet coral
Agaricia undata (Ellis and Solander, 1786)	A	scroll coral
Leptoseris cailleti (Duchassaing and Michelotti, 1864)	A	lace lettuce coral
Leptoseris cucullata (Ellis and Solander, 1786)	A	sunray lettuce coral
Siderastreidae		
Siderastrea radians (Pallas, 1766)	A	lesser starlet coral
Siderastrea siderea (Ellis and Solander, 1786)	A	massive starlet coral
Fungiidae		
Fungiacyathus crispus (Pourtalès, 1871)	A	
Poritidae		
Porites astreoides Lamarck, 1816	A	mustard hill coral
Porites branneri Rathbun, 1888	A	blue crust coral
Porites porites divaricata Lesueur, 1820	A	thin finger coral
Porites porites furcata Lamarck, 1816	A	branched finger coral
Porites porites porites (Pallas, 1766)	A	clubtip finger coral
SUBORDER FAVIINA		
Faviidae		
* *Colpophyllia natans* (Houttuyn, 1772)	A	boulder brain coral
Diploria clivosa (Ellis and Solander, 1786)	A	knobby brain coral
Diploria labyrinthiformis (Linnaeus, 1758)	A	grooved brain coral
Diploria strigosa (Dana, 1846)	A	symmetrical brain coral
* *Favia fragum* (Esper, 1793)	A	golfball coral
Favia gravida Verrill, 1868	A	
Manicina areolata areolata (Linnaeus, 1758)	A	rose coral
Manicina areolata mayori Wells, 1936	A	Tortugas rose coral
Montastraea annularis (Ellis and Solander, 1786)	A	boulder star coral
* *Montastraea cavernosa* (Linnaeus, 1767)	A	great star coral
* *Montastraea faveolata* (Ellis and Solander, 1786)	A	mountainous star coral
* *Montastraea franksi* (Gregory, 1895)	A	boulder star coral
* *Solenastrea bournoni* Milne Edwards and Haime, 1849	A	smooth star coral
Solenastrea hyades (Dana, 1846)	A	knobby star coral
***Rhizangiidae**		
* *Astrangia haimei* Verrill, 1866	P	
Astrangia poculata (Ellis and Solander, 1786)	A	northern star coral
* *Astrangia solitaria* (Lesueur, 1817)	A	dwarf cup coral
Oculinidae		
* *Madrepora carolina* (Pourtalès, 1871)	A	Pourtalès fan coral
* *Madrepora oculata* Linnaeus, 1758	P	
Oculina arbuscula Agassiz, 1864	A	compact ivory bush coral
Oculina diffusa Lamarck, 1816	A	diffuse ivory bush coral
* *Oculina profunda* Cairns, 1991	P	
Oculina robusta Pourtalès, 1871	A	robust ivory tree coral
Oculina tenella Pourtalès, 1871	A	delicate ivory bush coral
* *Oculina varicosa* Lesueur, 1831	A	fused ivory tree coral

SCIENTIFIC NAME	OCCURRENCE	COMMON NAME
Meandrinidae		
Dendrogyra cylindrus Ehrenberg, 1834	A	pillar coral
* *Dichocoenia stokesi* Milne Edwards and Haime, 1848	A	pineapple coral
Meandrina meandrites brasiliensis (Milne Edwards and Haime, 1848)	A	Brazilian rose coral
Meandrina meandrites meandrites (Linnaeus, 1758)	A	maze coral
Mussidae		
Isophyllastraea rigida (Dana, 1846)	A	rough star coral
* *Isophyllia sinuosa* (Ellis and Solander, 1786)	A	sinuous cactus coral
Mussa angulosa (Pallas, 1766)	A	large flower coral
Mycetophyllia aliciae Wells, 1973	A	knobby cactus coral
Mycetophyllia ferox Wells, 1973	A	rough cactus coral
* *Mycetophyllia lamarckiana* Milne Edwards and Haime, 1848	A	ridged cactus coral
Scolymia cubensis Milne Edwards and Haime, 1849	A	artichoke coral
Scolymia lacera (Pallas, 1766)	A	Atlantic mushroom coral
SUBORDER CARYOPHYLLIINA		
***Caryophylliidae**		
Anomocora fecunda (Pourtalès, 1871)	A	
* *Anomocora marchadi* (Chevalier, 1966)	A	
* *Anomocora prolifera* (Pourtalès, 1871)	A	
* *Caryophyllia alaskensis* Vaughan, 1941	P	
Caryophyllia ambrosia caribbeana Cairns, 1979	A	Caribbean horn coral
Caryophyllia arnoldi Vaughan, 1900	P	
* *Caryophyllia barbadensis* Cairns, 1979	A	
Caryophyllia berteriana Duchassaing, 1850	A	
Caryophyllia horologium Cairns, 1977	A	
* *Cladocora arbuscula* (Lesueur, 1820)	A	tube coral
* *Cladocora debilis* Milne Edwards and Haime, 1849	A	thin tube coral
Coenocyathus bowersi Vaughan, 1906	P	
* *Coenocyathus humanni* Cairns, 2000	A	ornate cup coral
* *Coenocyathus parvulus* (Cairns, 1979)	A	
Coenosmilia arbuscula Pourtalès, 1874	A	
* *Colangia immersa* Pourtalès, 1871	A	lesser speckled cup coral
Concentrotheca laevigata (Pourtalès, 1871)	A	
* *Crispatotrochus foxi* Durham and Barnard, 1952	P	
Dasmosmilia lymani (Pourtalès, 1871)	A	splitting cup coral
Dasmosmilia variegata (Pourtalès, 1871)	A	
Deltocyathus calcar (Pourtalès, 1874)	A	deep-sea star coral
Deltocyathus eccentricus Cairns, 1979	A	
* *Desmophyllum dianthus* (Esper, 1794)	A-P	cockscomb cup coral
Eusmilia fastigiata (Pallas, 1766)	A	smooth flower coral
* *Labyrinthocyathus quaylei* Durham, 1947	P	
* *Lophelia pertusa* (Linnaeus, 1758)	A-P	spider hazards
Nomlandia californica Durham and Barnard, 1952	P	
* *Oxysmilia rotundifolia* (Milne Edwards and Haime, 1848)	A	
Paracyathus montereyensis Durham, 1947	P	
Paracyathus pulchellus (Philippi, 1842)	A	papillose cup coral
* *Paracyathus stearnsii* Verrill, 1869	P	
* *Phacelocyathus flos* (Pourtalès, 1878)	A	two-tone cup coral

Occurrence abbreviations: A=Atlantic, P=Pacific, Ac=Arctic, E=Estuarine, F=Freshwater, I=Introduced.
*Change from first edition of 1991. +Not changed, but comment in Part II (Appendix 1).
^Species listed under both their hydroid and medusoid stages.

SCIENTIFIC NAME	OCCURRENCE	COMMON NAME
* *Phyllangia americana americana* Milne Edwards and Haime, 1849	A	hidden cup coral
* *Phyllangia pequegnatae* Cairns, 2000	A	
* *Polycyathus senegalensis* Chevalier, 1966	A	
Pourtalosmilia conferta Cairns, 1978	A	
* *Premocyathus cornuformis* (Pourtalès, 1868)	A	lesser horn coral
Rhizosmilia maculata (Pourtalès, 1874)	A	speckled cup coral
* *Thalamophyllia riisei* (Duchassaing and Michelotti, 1860)	A	baroque cave coral
Trochocyathus rawsonii Pourtalès, 1874	A	
***Turbinoliidae**		
* *Deltocyathoides stimpsonii* (Pourtalès, 1871)	A	
* *Sphenotrochus andrewianus moorei* Cairns, 2000	A	
Flabellidae		
* *Flabellum floridanum* Cairns, 1991	A	
Flabellum macandrewi Gray, 1849	A	splitting fan coral
* *Javania cailleti* (Duchassaing and Michelotti, 1864)	A-P	
* *Javania californica* Cairns, 1994	P	
Polymyces fragilis (Pourtalès, 1868)	A	twelve-root cup coral
* *Polymyces montereyensis* (Durham, 1947)	P	
Guyniidae		
+ *Guynia annulata* Duncan, 1872	A	
Schizocyathus fissilis Pourtalès, 1874	A	
+ *Stenocyathus vermiformis* (Pourtalès, 1868)	A	worm coral
***Gardineriidae**		
* *Gardineria paradoxa* (Pourtalès, 1868)	A	
SUBORDER DENDROPHYLLIINA		
***Dendrophylliidae**		
Balanophyllia elegans Verrill, 1864	P	
Balanophyllia floridana Pourtalès, 1868	A	porous cup coral
* *Balanophyllia palifera* Pourtalès, 1878	A	
* *Cladopsammia manuelensis* Chevalier, 1966	A	
* *Dendrophyllia oldroydae* Oldroyd, 1924	P	
* *Eguchipsammia cornucopia* Pourtalès, 1871	A	
* *Eguchipsammia gaditana* (Duncan, 1873)	A	
* *Eguchipsammia strigosa* Cairns, 2000	A	
* *Rhizopsammia goesi* (Lindström, 1877)	A	
* *Tubastraea coccinea* Lesson, 1829	A(I)	orange cup coral

PHYLUM CTENOPHORA—COMB JELLIES OR SEA WALNUTS

Class Tentaculata

ORDER CYDIPPIDA

Haeckeliidae

SCIENTIFIC NAME	OCCURRENCE	COMMON NAME
* *Aulacoctena acuminata* Mortensen, 1932	A-P	
* *Haeckelia beehleri* (Mayer, 1912)	A-P	
* *Haeckelia bimaculata* C. Carré and D. Carré, 1989	A-P	
* *Haeckelia rubra* (Kölliker, 1853)	A-P	

Occurrence abbreviations: A=Atlantic, P=Pacific, Ac=Arctic, E=Estuarine, F=Freshwater, I=Introduced.
*Change from first edition of 1991. +Not changed, but comment in Part II (Appendix 1).
^Species listed under both their hydroid and medusoid stages.

SCIENTIFIC NAME	OCCURRENCE	COMMON NAME
Bathyctenidae		
* *Bathyctena chuni* (Moser, 1909)	A-P	
Lampeidae		
Lampea lactea (Mayer, 1912)	A	
* *Lampea pancerina* (Chun, 1879)	A-P	
Pleurobrachiidae		
* *Hormiphora californensis* (Torrey, 1904)	P	
Hormiphora cucumis (Mertens, 1833)	P	
Hormiphora hormiphora (Gegenbaur, 1856)	A	
* *Pleurobrachia bachei* A. Agassiz, 1860	P	
Pleurobrachia brunnea Mayer, 1912	A	
Pleurobrachia pileus (O. F. Müller, 1776)	A	sea gooseberry
***Euplokamididae**		
Euplokamis crinita (Moser, 1909)	Ac	
* *Euplokamis dunlapae* Mills, 1987	A-P	
Mertensiidae		
* *Callianira bialata* delle Chiaje, 1841	A	
* *Charistephane fugiens* Chun, 1879	A-P	
Mertensia ovum (Fabricius, 1780)	A-P-Ac	
* Undescribed species (with pink-red tentacles)	P	
* Undescribed species "*Agmayeria tortugensis*"	A	Tortugas Red
Dryodoridae		
Dryodora glandiformis (Mertens, 1833)	A-P-Ac	
ORDER PLATYCTENIDA		
Coeloplanidae		
* *Coeloplana* sp. Smith, 1945	A	
* *Vallicula multiformis* Rankin, 1956	A-P(I)	
Tjalfiellidae		
Tjalfiella tristoma Mortensen, 1910	Ac	
ORDER THALASSOCALYCIDA		
Thalassocalycidae		
* *Thalassocalyce inconstans* Madin and Harbison, 1978	A-P	
ORDER LOBATA		
Bathocyroidae		
Bathocyroe fosteri Madin and Harbison, 1978	A-P	
Bolinopsidae		
* *Bolinopsis infundibulum* (O. F. Müller, 1776)	A-P-Ac	
Bolinopsis vitrea (L. Agassiz, 1860)	A	
Mnemiopsis gardeni L. Agassiz, 1860	A	
Mnemiopsis leidyi A. Agassiz, 1865	A	sea walnut
Mnemiopsis mccradyi Mayer, 1900	A	
Undescribed		
* Undescribed species "*Lampoctena sanguineventer*"	A-P	

Occurrence abbreviations: A=Atlantic, P=Pacific, Ac=Arctic, E=Estuarine, F=Freshwater, I=Introduced.
*Change from first edition of 1991. +Not changed, but comment in Part II (Appendix 1).
^Species listed under both their hydroid and medusoid stages.

SCIENTIFIC NAME	OCCURRENCE	COMMON NAME
Eurhamphaeidae		
* *Eurhamphaea vexilligera* Gegenbaur, 1856	A-P	
* *Deiopea kaloktenota* Chun, 1879	A-P	
* *Kiyohimea aurita* Komai and Tokioka, 1940	A-P	
* *Kiyohimea usagi* Matsumoto and Robison, 1992	P	
Leucotheidae		
Leucothea multicornis (Quoy and Gaimard, 1824)	A	
Leucothea ochracea Mayer, 1912	A	
* *Leucothea pulchra* Matsumoto, 1988	P	
Ocyropsidae		
* *Ocyropsis crystallina crystallina* (Rang, 1828)	A	
Ocyropsis crystallina guttata Harbison and Miller, 1986	A	
Ocyropsis maculata immaculata Harbison and Miller, 1986	A	
* *Ocyropsis maculata maculata* (Rang, 1828)	A-P	

ORDER CESTIDA

SCIENTIFIC NAME	OCCURRENCE	COMMON NAME
Cestidae		
Cestum veneris Lesueur, 1813	A-P	
* *Velamen parallelum* (Fol, 1869)	A-P	

Class Nuda

ORDER BEROIDA

SCIENTIFIC NAME	OCCURRENCE	COMMON NAME
Beroidae		
* *Beroe abyssicola* Mortensen, 1927	P	
Beroe cucumis Fabricius, 1780	A-P-Ac	
* *Beroe forskalii* Milne Edwards, 1841	A-P	
* *Beroe gracilis* Künne, 1939	P	
* *Beroe mitrata* (Moser, 1907)	P	
* *Beroe ovata* Bruguière, 1789	A	

Occurrence abbreviations: A=Atlantic, P=Pacific, Ac=Arctic, E=Estuarine, F=Freshwater, I=Introduced.
*Change from first edition of 1991. +Not changed, but comment in Part II (Appendix 1).
^Species listed under both their hydroid and medusoid stages.

PART II
Appendix 1:
Changes and Annotations to the 1991 Edition

The comments and explanatory notes below are keyed to the appropriate scientific name as indicated by the asterisk (*) or plus sign (+) in the main list, Part I. An asterisk signals some change from the first edition; a plus sign indicates that no change has been made but nonetheless a comment appears in Part II. Entries are in the same order as in the list and are grouped, for convenience, by page. References are listed in Part III, grouped by higher taxonomic group.

Page 9

Carybdea marsupialis. Material called *Carybdea rastoni* collected in Santa Barbara, California, is probably all *C. marsupialis*, as identified by Larson and Arneson (1990). Use of the (valid) name *C. rastonii* for this California material still occurs occasionally (see Matsumoto 1995).

Haliclystus. Hirano (1986 and 1997) has done careful morphological comparisons of Northern Hemisphere species of *Haliclystus*, which is the authority used for species identifications here.

Haliclystus auricula. Although the name has been used much more widely, according to Hirano (1997), the original species described by Clark (1878) appears to be restricted to the North Atlantic (on both European and North American coasts).

Haliclystus octoradiatus. This name has recently been used for specimens on the Pacific coast of North America (Eckelbarger and Larson 1993), but Hirano (1997) suggests that *H. octoradiatus* may be restricted to the European North Atlantic. If further studies confirm this, *H. octoradiatus* should be deleted from this list in the future.

Haliclystus salpinx. No changes are proposed in the identification of this species, which has a peculiarly disjunct distribution, occurring in the North Atlantic as well as two localities in the San Juan Islands, Washington State.

Haliclystus stejnegeri. Hirano (1986, 1997) suggests that this species is boreal, occurring from Alaska to northern Japan. Use of this species name for animals in southern British Columbia and Washington State is probably incorrect.

Haliclystus sp. undescribed "*californiensis.*" Gwilliam (1956) suggested that a single specimen from Southern California was sufficiently distinct to merit its own species. Gwilliam described it quite thoroughly in his Ph.D. dissertation (1956), but that description was never formally published. The species has seemingly not been collected since that initial observation.

Haliclystus sp. undescribed "*sanjuanensis.*" According to Hirano (1997), this seems to be a distinct morphological species, known from many locations in southern British Columbia and northwestern Washington State. It has never been formally described, but the *nomen nudem H. sanjuanensis* appears several times in the literature (e.g., Hyman 1940); it has also variously been called *H. auricula, H. octoradiatus, H. stejnegeri,* and *H. sanjuanensis.*

Undescribed species "*Stenoscyphopsis vermiformis.*" Gwilliam (1956) collected this species on nearshore kelp in California and described it quite thoroughly in his Ph.D. dissertation, but it was not formally published. The species has not been collected since that initial observation.

Kyopodiidae. The family name (Larson 1988) was misspelled in the first edition. The family was also inadvertently moved into the suborder Cleistocarpida, in the first edition, but was originally described as a member of the suborder Eleutherocarpida (Larson 1988) and should remain there.

Thaumatoscyphinae. This subfamily appears to be in common use, containing at present only the genus *Manania* (Larson and Fautin 1989).

Manania auricula. The species is still valid (Larson and Fautin 1989) but was omitted from the first edition; it was reported from Maine (Clark 1863).

Craterolophinae. This was omitted from the first edition but was in use by Kramp (1961) for the genus *Craterolophus.*

Craterolophus convolvulus. The placement of *Craterolophus* in the Thaumatoscyphinae in the first edition seems to have been an error, since Larson and Fautin (1989) state that the subfamily Thaumatoscyphinae contains only the genus *Manania.*

Page 10

Atorellidae. The family was omitted from the first edition because Mills et al. (1987) moved the genus *Atorella* into the Nausithoidae, but subsequent authors, including Mianzan and Cornelius (1999), retain this monotypic family for *Atorella.*

Atorella octogonos. The species name was spelled incorrectly in the first edition.

Linuche unguiculata. Author Swartz was spelled incorrectly in the first edition (and in Kramp 1961).

Paraphyllina intermedia. Collected by C. E. Mills and others off California in deep tows, sometimes with open nets, this species probably lives below the depth range covered in this manual, but the precise depth range has not been established.

Periphyllopsis galathea. Collected in deep tows with open nets by B. J. Burd and R. E. Thomson off British Columbia and identified by C. E. Mills, this species probably lives below the depth range covered in this manual, but the precise depth range has not been established.

Tetraplatidae. This monospecific family containing a single highly modified medusa is here included in the Coronatae, as suggested by Ralph (1960) and followed by Wrobel and Mills (1998).

Tetraplatia volitans. This cosmopolitan species is reported along the Pacific coast by Wrobel and Mills (1998); it has been collected several times in central California.

Chrysaora achlyos. This recently described, very large species occurs only occasionally nearshore and beached in Southern California (Martin et al. 1997); strandings there are known from 1989 and 1999.

Chrysaora melanaster. This species is known from the Pacific Ocean in the Bering Sea (Brodeur et al. 1999) and Prince William Sound (photos shown to C. E. Mills 2000).

Drymonema dalmatinum. Author Haeckel was misspelled in the first edition. This is considered to be an introduced species (C. Mills 2002, personal communication).

Ulmaridae. The subfamilies Aureliinae, Sthenoniinae, Poraliinae, Stygiomedusinae, and Deepstariinae are typically used within this family (Kramp 1961; Larson 1986, Wrobel and Mills 1998) and thus have been added to this edition.

Aureliinae. See Ulmaridae.

Aurelia aurita. This species or species-complex has nearly worldwide distribution, including the American Atlantic seaboard, the Gulf of Mexico, and a few specific locations along the American Pacific coast. It is distinguished by four oral arms that meet in the center of the subumbrella without a large central manubrial mass; it carries developing larvae on the innermost edges of the oral arms (Wrobel and Mills 1998). It is on display at many public aquariums in America and seems to be present in San Francisco Bay (Greenberg et al. 1996). This species is considered to be introduced to the Pacific Coast in lagoons in San Francisco (Rees and Gershwin 2000).

Aurelia labiata. This species, which was omitted from the first edition, seems to occur primarily on the west coast of North America. It is distinguished by four oral arms that are generally shorter than the bell and separated at the center of the subumbrella by a large, frilly, conical gelatinous manubrium; it carries developing larvae on the conical central manubrium (Wrobel and Mills 1998). L. A. Gershwin (at this writing, a student at the University of California, Berkeley) first noted the distinction between this species and the *A. "aurita"* morph in aquariums along the Pacific coast where both forms are often exhibited in the same tank.

Aurelia limbata. This species was originally described as an *Aurelia* (although Brandt redescribed it three years later as *Diplocraspedon limbata*), so the parentheses around the data and author in the first edition were in error.

Sthenoniinae. See Ulmaridae.

Phacellophora camtschatica. The common name "fried egg jellyfish" is used about as often as "egg yolk jelly" for this species.

Poraliinae. See Ulmaridae.

Stygiomedusinae. See Ulmaridae.

Stygiomedusa gigantea. Reported by Wrobel and Mills (1998) from deep trawls off California and Washington. Pieces of this species were found in open trawls collecting to maximum depths of 800 to 1,700 m off Southern California; it probably lives below the depth range covered in this manual, but the precise depth range has not been established.

Deepstariinae. See Ulmaridae.

Page 11

Phyllorhiza punctata. This species is now considered to be introduced (C. Mils 2002, personal communication).

Rhopilema verrilli. The common name has been added in this edition.

Stomolophus meleagris. Two common names, "cabbagehead" and "cannonball jellyfish," are used for this species. It is also known as "jellyball" in some areas of the southeastern United States.

Subclass Leptolida. The name of this subclass follows the classification of Cornelius (1992 1995a).

Order Anthoathecatae. This order was spelled "Athecatae" in the first edition. Cornelius (1992) proposed the term Anthoathecata (emended to Anthoathecatae by Cornelius 1995a) as a single replacement name for the taxon previously called Athecatae (=Athecata =Gymnoblastea), or Anthomedusae, or some cumbersome combination of the two. These names are all vestiges of separate classification systems that had developed in the past for hydroids and hydromedusae. Arguments have been made that the term "Anthomedusae" be retained and applied to the group, but we consider this confusing and unjustified when a name more inclusive and representative of both polypoid and medusoid stages is available. In terms of phylogeny, relationships within taxa of the order are poorly known; only the family Stylasteridae has been subjected to detailed cladistic analysis (Cairns 1984). The provisional classification employed here largely follows that adopted by Calder (1988) for filiferan families (Clavidae through Eudendriidae), and by Brinckmann-Voss (1970) and Petersen (1990) for capitate families (Moerisiidae through Pennariidae). Major synonyms used as valid names by Fraser (1937, 1944), whose two comprehensive publications are still the primary guidebooks to the hydroids of Canada and the United States, are included in this edition.

Clavidae. *Turris neglecta* Lesson, 1843, a presumed clavid reported from the Queen Charlotte Islands (Fraser 1937), is a *nomen dubium* (Russell 1953; Calder 1988) and has been excluded from this list, as in the first edition. There is evidence from molecular studies that the genus *Clava* should be assigned to the same family as *Hydractinia* and related genera. It is likely that the taxonomic scope and name of the taxon currently called Clavidae will change in the immediate future. But, until this issue is clearly resolved, we retain current usage.

Clava multicornis. *Clava leptostyla* L. Agassiz, 1862 is regarded as a junior subjective synonym (see Edwards and Harvey 1975).

Cordylophora caspia. We add this species to the Pacific coast list in this edition (see Mills and Miller 1987). *Cordylophora lacustris* Allman, 1844 is regarded as a junior subjective synonym (see Vervoort 1964).

Merona cornucopiae. This appeared as *Tubiclava cornucopiae* in the first edition. *Merona* differs from *Tubiclava* and other clavid genera in having both nematothecae and polymorphic colonies with gonophores on gonozooids (Millard 1975). Based on its morphology, this species is referable to the former.

Merona laxa. This species was given as *Tubiclava laxa* in the first edition. See *Merona cornucopiae* above.

Rhizogeton ezoense. We add this to the species listed in this edition (see Brinckmann-Voss 1996).

Rhizogeton fusiforme. The specific name was misspelled *fusiformis* in the first edition.

Rhizogeton nematophorum. We add this to the species listed in this edition (see Brinckmann-Voss 1996).

Rhizogeton nudum. The specific name was misspelled in the first edition. We add this species to the Pacific coast list in this edition (see Brinckmann-Voss,1996).

Turritopsis nutricula. The date was incorrectly cited as 1859 in the first edition. McCrady's account of this species is now known to have been published in 1857 (Calder, Stephens, and Sanders 1992).

Hydractiniidae. This family is in need of revision. A seemingly strong but as yet little-studied affinity exists between hydractiniids and clavids, as well as bougainvilliids. Moreover, the generic classification of Hydractiniidae is in a state of flux. Until more is known about the taxonomy of the group, *Podocoryna* (*Podocoryne* is an incorrect subsequent spelling) and *Stylactaria* are included in the synonymy of *Hydractinia* (see Boero et al. 1998).

Hydractinia americana. We add this to the species listed in this edition. It was given as *Podocoryna carnea* (M. Sars, 1846) in the first edition; that binomen is deleted here. The opinion of Edwards (1972) that the North American population belongs to a different species from *Podocoryna carnea* of Europe is adopted. See Hydractiniidae above.

Hydractinia arge. This was given as *Stylactaria arge* in the first edition. See Hydractiniidae above.

Hydractinia carica. *Hydractinia minuta* Bonnevie, 1898 is regarded as a junior subjective synonym (see Broch 1916).

Hydractinia echinata. The date was incorrectly cited as 1823 in the first edition.

Hydractinia echinata, *H. polyclina*, *H. symbiolongicarpus*, and *H. symbiopollicaris* of the northeastern coast of the United States are considered to be a sibling species group, distinguishable at the molecular but not at the morphological level (Buss and Yund 1989). Of this group, only *H. echinata* was included in the first edition.

Hydractinia hooperii. We add this to the species listed in this edition. It was regarded as conspecific with *Hydractinia arge* and not included in the first edition, but is now thought to differ in its mode of reproduction (see Namikawa 1991).

Hydractinia polyclina. We add this to the species listed in this edition. It was regarded as conspecific with *Hydractinia echinata* and not included in the first edition; see *H. echinata* above.

Hydractinia selena. This was given as *Podocoryna selena* in the first edition. See Hydractiniidae above.

Hydractinia symbiolongicarpus. We add this to the species listed in this edition. It was regarded as conspecific with *Hydractinia echinata* and not included in the first edition; see *H. echinata* above.

Hydractinia symbiopollicaris. We add this to the species listed in this edition. It was regarded as conspecific with *Hydractinia echinata* and not included in the first edition; see *H. echinata* above.

Page 12

Stylaster erubescens erubescens. Zibrowius and Cairns (1992) described three additional subspecies of *S. erubescens* from the northern and eastern Atlantic, making it necessary to specify that the western Atlantic subspecies is the nominate one.

Rhysiidae. We add this to the families listed in this edition (see Brinckmann-Voss, Lickey, and Mills 1993).

Rhysia fletcheri. We add this to the species listed in this edition (see Brinckmann-Voss, Lickey, and Mills 1993).

Bougainvilliidae. Two nominal species of bougainvilliids—*Bougainvillia longicirra* Stechow, 1914 and *Bougainvillia robusta* (Fraser, 1938)—listed in the first edition, having Caribbean distributions only, have been deleted here. The name Atractylidae Hincks, 1868, applied to this family by Fraser (1937, 1944), is a junior subjective synonym of Bougainvilliidae Lütken, 1850.

Bimeria vestita. Bimeria humilis Allman, 1877, listed as a valid species in the first edition, is regarded as junior subjective synonym (see Calder 1988).

Bougainvillia muscus. We add this to the species listed in this edition. *Bougainvillia ramosa* (sensu van Beneden, 1844) is regarded as a junior subjective synonym (see Calder 1988).

Bougainvillia superciliaris. Bougainvillia mertensi L. Agassiz, 1862 is regarded as a junior subjective synonym (see Vannucci and Rees 1961).

Dicoryne flexuosa. The date was incorrectly cited as 1873 in the first edition.

Garveia franciscana. Bimeria tunicata Fraser, 1943 is regarded as a junior subjective synonym (see Vervoort 1964). *Calyptospadix cerulea* Clarke, 1882 may also prove conspecific.

Rhizorhagium roseum. Garveia groenlandica Levinsen, 1893 is regarded as a junior subjective synonym (see Rees 1938). *Perigonimus formosus* Fewkes, 1889 from California seems similar to this species (Rees 1956).

Amphinema dinema. Perigonimus serpens Allman, 1863 is regarded as a junior subjective synonym (see Rees 1956). Fraser's (1937) report of this hydroid from the Queen Charlotte Islands, as *Perigonimus serpens*, is considered doubtful because the medusa of *Amphinema dinema* has not been reported from the Pacific Northwest.

Hydrichthys mirus. As *Stomotoca mira* in the first edition (see Boero, Bouillon, and Gravili 1991).

Leuckartiara octona. Perigonimus jonesii Osburn and Hargitt, 1894, *Perigonimus repens* (Wright, 1857), *Perigonimus sessilis* (Wright, 1857), and *Perigonimus pugetensis* Heath, 1910 are regarded as junior subjective synonyms (see Rees 1956).

Page 13

Halimedusa typus. The hydroid stage is added to the list in this edition (Mills 2000).

Eudendriidae. Major revisions may be expected in the taxonomy of this family. Many nominal species of *Eudendrium* have been founded in the past on fragmentary specimens and distinguished at least in part on the basis of unreliable characters. Several names below are based originally on material that even lacked hydranths and gonophores, generally considered necessary for identification. Although such nominal species in fact represent nomina dubia, most are retained here because a comprehensive revision of the genus has yet to be published. *Eudendrium tenellum* Allman, 1877 is clearly beyond identification (Calder, 1988) and has been excluded, as in the first edition.

Eudendrium album. The date was incorrectly cited as 1898 in the first edition. The specific name was made available earlier by Nutting (1896).

Eudendrium arbuscula. The specific name was misspelled *arbusculum* in the first edition.

Eudendrium capillare. Eudendrium tenue A. Agassiz, 1865, listed as a valid species in the first edition, is regarded as a junior subjective synonym (see Calder 1988).

Eudendrium vaginatum. Eudendrium annulatum Norman, 1864, listed as a valid species in the first edition, is regarded as a junior synonym (Sheiko and Stepanjants 1997).

Myrionema amboinense. Eudendrium hargitti Congdon, 1906 is regarded as a junior subjective synonym (see Calder 1988).

Monobrachiidae. This family is maintained as valid and retained in the Anthoathecatae, although Bouillon (1995) recently assigned the genus *Monobrachium* to the Olindiidae, Order Limnomedusae.

Proboscidactylidae. Bouillon (1995) assigned this family to the order Limnomedusae, but Schuchert (1996) and others have included it in the Anthoathecatae.

Moerisia lyonsi. The specific identity of the leptolid in North America that has frequently been assigned this binomen is uncertain; Petersen (1990: Fig. 5B) referred it instead to *Moerisia gangetica* Kramp, 1958.

Hydridae. This family was inadvertently placed in Order Anthomedusae in the first edition.

Page 14

Protohydridae. This family was inadvertently placed in Order Anthomedusae in the first edition.

Protohydra leuckarti. We add this species to the Pacific coast list in this edition (see Mills and Miller 1987).

Cladocoryne floccosa. We add this species to the Pacific coast list in this edition (see Lees 1968). *Cladocoryne pelagica* Allman, 1876 is regarded as a junior subjective synonym (see Hirohito 1988).

Zanclea bomala. We add this to the species listed in this edition (see Boero et al. 2000).

Zanclea gemmosa. We add this to the species listed in this edition. It was regarded as conspecific with *Zanclea costata* and not included in the first edition. The taxonomy of this difficult genus needs further revision, and it seems advisable for now to provisionally include this nominal species.

Zanclella bryozoophila. We add this to the species listed in this edition (see Boero and Hewitt 1992).

Hataia parva. We add this to the species listed in this edition (see Brinckmann-Voss 1996).

Corymorphidae. As in the first edition, *Euphysa* is included in the Corymorphidae, following Rees (1957), Brinckmann-Voss (1970), Petersen (1990), and others. Some authors (e.g., Bouillon 1995) recognize a separate family Euphysidae for this and related genera. *Zyzzyzus* has been reassigned in this edition from Corymorphidae to Tubulariidae (q.v.).

Corymorpha bigelowi. Reported in genus *Euphysora* in the first edition.

Corymorpha groenlandica. Rhizonema carnea Clark, 1877 from Alaska may be conspecific (Calder and Stephens 1997).

Corymorpha pendula. Tubularia crassa Fraser, 1941 is regarded as a junior subjective synonym (see Calder 1975).

Euphysa farcta. This may prove to be the hydroid of the medusa *Euphysa aurata* Forbes, 1848.

Euphysa peregrina. This was listed as *Hypolytus peregrinus* in the first edition. Petersen (1990) has been followed here in regarding *Hypolytus* Murbach, 1899 as congeneric with *Euphysa* Forbes, 1848.

Tubulariidae. Petersen (1990) extensively revised this family. His redefinitions of the tubulariid genera *Tubularia*, *Ectopleura*, and *Hybocodon* are provisionally followed here. Too, the genus *Zyzzyzus* is transferred from the Corymorphidae to the Tubulariidae in this edition, following his classification. One species listed in the first edition (*Tubularia multitentaculata* Fraser, 1938, from Baja California) has been deleted because its known range is outside the area of coverage.

Ectopleura crocea. This was listed as *Tubularia crocea* in the first edition (see Petersen 1990).

Ectopleura dumortierii. The specific name was misspelled *dumortieri* in the first edition; the correct original spelling is *dumortierii. Tubularia cristata* McCrady, 1859 is regarded as a junior subjective synonym (see Calder 1975); so too is *Ectopleura prolifica* Hargitt, 1908, which was included as a valid species in the first edition (see Petersen 1990).

Ectopleura larynx. This was listed as *Tubularia larynx* in the first edition (see Petersen 1990). *Tubularia spectabilis* (L. Agassiz, 1862) and *Tubularia tenella* (L. Agassiz, 1862) are regarded as junior subjective synonyms (see Vervoort 1946).

Ectopleura marina. This was listed as *Tubularia marina* in the first edition (see Petersen 1990).

Tubularia indivisa. Tubularia couthouyi L. Agassiz, 1862, included as a valid species in the first edition, is regarded as a junior subjective synonym (see Petersen 1990).

Tubularia regalis. Tubularia borealis Clark, 1876, included as a valid species in the first edition, is regarded as a junior subjective synonym (see Petersen 1990).

Zyzzyzus floridanus. This was assigned to Corymorphidae instead of Tubulariidae in the first edition (see Petersen 1990).

Margelopsis gibbesii. The specific name was misspelled *gibbesi* in the first edition; the correct original spelling is *gibbesii.*

Page 15

Candelabridae. Fraser (1944) applied the name Myriothelidae Hincks, 1868, to this family. The substitute name Candelabridae Stechow, 1921 seems to be in prevailing usage and is taken as valid under provisions of the International Code of Zoological Nomenclature (Art. 40).

Candelabrum phrygium. Blastothela rosea Verrill, 1878 is likely a junior subjective synonym.

Corynidae. The generic revisions within Corynidae by Petersen (1990) are generally adopted in this edition. In the first edition, *Sphaerocoryne agassizii* (McCrady, 1859) was assigned to this family; that species is referred here to Sphaerocorynidae (q.v.).

Coryne cliffordi. We add this to the species listed in this edition (see Brinckmann-Voss 1989).

Coryne eximia. This was listed as *Sarsia eximia* in the first edition (see Petersen 1990).

Coryne hincksii. The specific name was misspelled *hincksi* in the first edition; the correct original spelling is *hincksii.*

Coryne japonica. We add this to the species listed in this edition (see Mills and Miller 1987).

Coryne producta. We add this to the species listed in this edition (see Brinckmann-Voss 1996).

Coryne pusilla. The author and date were mistakenly enclosed in parentheses in the first edition.

Coryne vermicularis Hincks, 1866 is regarded as a junior subjective synonym (see Broch 1916).

Sarsia bella. We add this to the species listed in this edition (Brinckmann-Voss 2000).

Sarsia tubulosa. Sarsia mirabilis L. Agassiz, 1849 has commonly been considered conspecific with *Sarsia tubulosa*, but the identity of the hydroids attributed to *S. mirabilis* by Agassiz (1862) needs to be clarified by further life cycle studies. Fraser's

(1937) report of the hydroid of *Syncoryne mirabilis* from the Pacific coast requires verification.

Sphaerocorynidae. We add this to the families listed in this edition; for a discussion of Sphaerocorynidae, see Calder (1988) and Petersen (1990).

Sphaerocoryne agassizii. Assigned to family Corynidae in the first edition, this species is included in Sphaerocorynidae here (see Calder 1988; Petersen 1990).

Cladonema radiatum. Cladomena mayeri Perkins, 1906 is regarded as a junior subjective synonym (see Calder 1988).

Pennariidae. We add this to the families listed in this edition; the name Halocordylidae was applied to this family in the first edition. With the type genus *Halocordyle* Allman, 1872 now regarded as a junior subjective synonym of *Pennaria* Goldfuss, 1820 (see *Pennaria disticha* below), and with the family name Pennariidae McCrady, 1859 predating Halocordylidae Stechow, 1921 and regaining prevailing usage at the end of the 20th century, Pennariidae is recognized as valid here.

Pennaria disticha. This was listed as *Halocordyle disticha* in the first edition. Gibbons and Ryland (1989) presented evidence that *Pennaria* is the valid name of the genus rather than *Halocordyle. Pennaria tiarella* Ayres, 1854 is regarded as a junior subjective synonym (see Calder 1988).

Order Anthoathecatae. See page 48.

Turritopsis nutricula. We add this to the Pacific coast list in this edition (see Alvariño 1999). This species is now considered to be introduced on both east and west coasts (C. Mills 2002, personal communication)

Family Hydractiniidae. The generic names of *Podocoryna* and *Stylactaria,* used in the first edition, are changed to *Hydractinia* in this edition. See page 49.

Hydractinia americana. See page 49.

Hydractinia arge. See page 49.

Hydractinia dubia. See Hydractiniidae, page 49.

Hydractinia hooperi. We add this to the species listed in this edition. The medusa is short lived.

Hydractinia minima. See Hydractiniidae, page 49.

Hydractinia minuta. See Hydractiniidae, page 49.

Hydractinia selena. The medusa is not reported from the field but raised from field-collected hydroids in the laboratory.

Page 16

Bougainvillia muscus. We add this to the Pacific coast list in this edition. It was reported by Mills (in Kozloff 1996) as *Bougainvillia ramosa* but corrected to *Bougainvilla muscus* in "Additions and corrections" of the same book.

Bougainvillia principis. This name replaces *Bougainvillia multitentaculata* of the first edition (see Mills in Kozloff 1996).

Rathkea octopunctata. Although the medusa is very abundant in spring, the hydroid has not been reported in A-P-Ac, probably because of its small size. It has been reported for the North Sea (Werner 1985).

Amphinema australis. This species is not present off Vancouver Island as reported in Boero et al. (1997), being cited wrongly from Foerster (1923).

Amphinema dinema. We add this to the Pacific coast list because medusae have been raised from hydroids from California; however, systematic relation to *A. rugosum* needs to be verified (Rees 2000a)

Amphinema platyhedos. The publication year is corrected from 1985 to 1983 (Arai and Brinckmann-Voss 1983).

Amphinema turrida. We add this species to the Pacific coast list because the picture given by Wrobel and Mills (1998:26) as *Amphinema platyhedos* is definitely not that species but more likely *Amphinema turrida.*

Annatira affinis. We add this to the species listed in this edition (Wrobel and Mills 1998).

Catablema nodulosa. The specific name was misspelled *C. nodulosum* in the first edition (see Arai and Brinckmann-Voss 1980).

Larsonia pterophylla. This species is transferred from *Stomotoca* to genus *Larsonia* by Boeroet al. (1991).

Leuckartiara zacae. We add this species to the Pacific coast list in this edition (Alvariño 1999 and Brinckmann-Voss, private collections for Strait of Juan de Fuca, deposited at ROM).

Merga reesi. We add this species to the Pacific coast list in this edition. Although outside the 200 m depth limit, it was within 320 km of the Canadian coast. (Brinckmann-Voss and Arai 1998).

Neoturris fontata. The record for this species is doubtful; it was reported once only (Bigelow 1909).

Pandea rubra. This species is newly recorded within 320 km of the Canadian coast (Brinckmann-Voss and Arai 1998).

Stomotoca atra. The hydroid has been described from the Indo-Pacific but not from North America (Boero and Bouillon 1989).

Halimedusa typus. Additional information on distribution (Brinckmann-Voss and Arai 1998) and hydroid (Mills 2000) has been gathered since the first edition.

Calycopsidae. *Meator rubater* is deleted from this edition. Its inclusion in the first edition was based on its presence in Bering Sea, now interpreted to be the Asiatic coast of the Bering Sea.

Bythotiara depressa. New distributional records and comparisons to *Heterotiara annonyma* are made by Brinckmann-Voss and Arai (1998).

Page 17

Calycopsis bigelowi. We add this species to the Pacific coast list in this edition; it was recorded outside the continental shelf but within 320 km (Brinckmann-Voss and Arai 1989). *Calycopsis simulans* is deleted in this edition because its taxonomic status is doubtful.

Eumedusa birulai. This species is restricted to the Arctic in this edition.

Heterotiara annonyma. We add this species to the Pacific coast list in this edition (Alvariño 1999), California only (see discussion of *Bythotiara depressa).*

Moerisia lyonsi. See page 50. This species is now considered to be introduced in Atlantic estuaries (C. Mills 2002, personal communication).

Polyorchis penicillatus. The species name was *penicillata* in the first edition.

Zancleopsidae. See *Zancleopsis dichotoma.*

Zancleopsis dichotoma. We place this species in the Zancleopsidae, established by Bouillon (1978), and not in the Zancleidae, as in the first edition.

Paragotoeidae. This family is new in this edition.

Paragotoea bathybia. We add this species to the Pacific coast list in this edition (Brinckmann-Voss and Arai 1998).

Family incertae sedis (suborder Sphaerocorynida). Petersen (1990) placed the genus *Euphysilla* as "fam.inc.sedis" in the suborder Sphaerocorynida; the genus was therefore removed in this edition from the Corymorphidae (first edition) to family incertae sedis, suborder Sphaerocorynida.

Zanclea bomala. The medusa is added to this edition, but it is only known from laboratory rearing (see reference to hydroid literature).

Zanclea costata. The record for the Pacific is deleted in this edition. Neither *Zanclea costata* nor *Z. gemmosa* was reported for U.S. or Canadian Pacific waters in their medusa stage.

Zanclea gemmosa. We add this to the species listed in this edition.

Zanclella bryozoophila. The medusa is added to this edition, but it is known only from laboratory rearing (see reference to hydroid literature).

Family Corymorphidae. The generic status of some species was changed by Petersen (1990), which is followed in this edition.

Euphysa aurata. The hydroid is known (Brinckmann-Voss 1970) but not reported for the United States and Canada.

Euphysa vervoorti. We add this to the species listed in this edition (Brinckmann-Voss and Arai 1998).

Corymorpha bigelowi, Corymorpha gracilis, Corymorpha forbesi. The genus is changed to *Corymorpha* in this edition from other genera (e.g., *Euphysora* and *Vannuccia*) in the first edition (see Petersen 1990).

Corymorpha nutans. We add this species to the Pacific coast list in this edition (Alvariño 1999).

Tubulariidae. *Climacodon ikarii* is deleted in this edition.

Ectopleura dumortierii. The specific name was misspelled *dumortieri* in the first edition.

Margelopsidae. The family name is added in this edition.

Margelopsis gibbesii. The specific name was misspelled *gibbesi* in the first edition.

Page 18

Corynidae. Genera of this family were rearranged by Petersen (1990) since the first edition, which is followed in this edition (see also page 51) although some of his generic characters need to be redefined. See discussion in Brinckmann-Voss (2000). *Sarsia occidentalis* deleted in this edition, because poorly defined species (see Arai and Brinckmann-Voss 1980). *Sarsia reticulata* deleted in this edition because of uncertain taxonomic status.

Coryne cliffordi. We add this to the species listed in this edition (Brinckmann-Voss 1989).

Coryne eximia. The genus name is changed in this edition (Petersen 1990), and the species is new to the medusoid list.

Coryne japonica. This species has been transferred from genus *Sarsia.*

Dicodonium floridana. This is a poorly defined genus; most of its species will need to be placed in different genera or even families in the future (see Petersen 1990; Brinckmann-Voss and Arai 1998).

Sarsia bella. We add this to the species listed in this edition (Brinckmann-Voss 2000).

Sarsia occulta. The medusa is added to this edition, but it is known only from laboratory rearing (Brinckmann-Voss, work in progress).

Sphaerocorynidae. We add this to the families listed in this edition.

Sphaerocoryne agassizii. This species is moved from the Corynidae in the first edition to the Sphaerocorynidae in this edition.

Cladonema radiatum. This species is considered to be introduced on the west and east coasts (C. Mills 2002, personal communication).

Pennariidae. This family replaces the Halocordylidae in this edition.

Pennaria disticha. This family replaces the *Halycordyle disticha* in this edition. See page 52.

Family incertae sedis (suborder Tubulariida). We add this family in this edition.

Plotocnide borealis. This species is moved from family Corymorphidae (in the first edition) to family incertae sedis (in this this edition) because of its uncertain position within any of the existing families (see Petersen 1990; Schuchert 1996).

Family incertae sedis (order Anthoathecatae - Anthomedusae). We add this family in this edition.

Paulinum lineatum. We add this to the species listed in this edition. Superficially, it resembles *Velella* (Family Porpitidae) (see hydroid list); however, it is distinct from *Velella* and related genera through morphology of manubrium, marginal bulbs, and cnidocysts.

Trichydridae. We add this family in this edition.

Trichydra pudica. We add this to the species listed in this edition (see Arai and Brinckmann-Voss 1980).

Craspedacusta sowerbyi. The home range of this species is apparently the Yangtze Kiang Valley, China, where it is called the "peach blossom fish" or "peach blossom fan" because the medusa stage is most common when peach trees are in full bloom. As early as 1880, it was introduced to Regents Park, London, as an associate of aquatic plants that were impounded there. Since then, it has spread throughout Europe, Asia, and North and South America and is widely reported from rivers and lakes in almost every state of the United States between 33° and 43° N. It is particularly common in manmade lakes, where the medusa pulsates on the surface usually in late summer but sometimes as early as July 1. Useful references to this peculiar limnomedusan are Acker (1976) and Bushnell and Porter (1967).

Maeotias marginata (Modeer 1791). This name has priority over *M. inexpectata* Ostraumoff, 1886 (in the first edition), which is a junior synonym of *M. marginata*. The species was recently moved from the Moerisiidae, Order Anthoathecatae to the family Olindiidae, Order Limnomedusae because of the presence of statocysts (Mills and Rees 2000). This species is now considered to be introduced in low salinity on both east and west coasts (Mills and Rees 2000).

Order Leptothecatae. This was given as Order Thecatae in the first edition. Cornelius (1992) founded Leptothecata (emended to Leptothecatae by Cornelius 1995a) as a replacement name for the taxon previously called Thecatae (=Thecata =Calyptoblastea), or Leptomedusae, or some combination of the two. See Order Anthoathecatae, above. The arrangement of leptothecate families in this edition is modified from that of Bouillon (1995), with plumularioids (Kirchenpaueriidae, Plumulariidae, Halopterididae, Aglaopheniidae) listed immediately after haleciids to reflect the relationships between the two. The sequence of plumularioid families adopted here follows Calder (1997). This classification is provisional because no overall phylogenetic analysis of the group has been undertaken. As noted for anthoathecates above, names in Fraser's (1937, 1944) books that are now considered synonyms of binomena listed herein are noted here.

Melicertidae. We add this family in this edition.

Melicertum octocostatum. The hydroid stage is added in this edition (Brinckmann-Voss 1980).

Page 19

Aequorea forskalea. *Aequorea aequorea* (Forskål, 1775) in the first edition, that name is preoccupied (Cornelius 1995a).

Blackfordia virginica. We add this species to the Pacific coast list in this edition as esturine introduction (Mills and Rees 2000).

Opercularella rugosa. This species was given as *Campanulina rugosa* (family Campanulinidae) in the first edition, but it cannot be assigned to the genus *Campanulina* or to the family based on it (see Rees 1939; Calder 1991).

Calycella syringa. *Lafoea pygmaea* Hincks, 1868 is regarded as a junior subjective synonym (see Cornelius 1995a).

Family incertae sedis. The polyphyletic assemblage of taxa included under this heading is often incorrectly placed in the family Campanulinidae (see Rees 1939; Calder 1991).

Laodiceidae. Two of the species (*Modeeria rotunda*, *Stegopoma plicatile*) assigned to this family in the first edition have been referred to the Tiarannidae (q.v.).

Tiarannidae. We add this to the list of families in the hydroid section of this edition. The two species included in this taxon here (*Modeeria rotunda*; *Stegopoma plicatile*) were listed under Laodiceidae in the first edition (see Cornelius 1995a).

Modeeria rotunda. *Stegopoma fastigiata* (Alder 1860) is regarded as a junior subjective synonym (see Edwards 1973).

Eucheilotidae. See Cirrholoveniidae, page 60.

Eucheilota bakeri. This species was included by Nutting (1915) and Fraser (1937) as *Clytia bakeri*, a campanulariid; instead, its medusa is a referable to *Eucheilota*, a lovenellid (see Kramp 1961).

Lovenella nodosa. We add this to the species listed in this edition (see Lees 1986).

Lovenella producta. The date was incorrectly cited as 1873 in the first edition.

Halecium articulosum. This species was regarded as questionably conspecific with *Halecium sessile* by Cornelius (1995a).

Halecium delicatulum. We add this to the species listed in this edition. Two junior subjective synonyms (*Halecium flexile* Allman, 1888; *Halecium parvulum* Bale, 1888; see Rees and Vervoort 1987) were included as valid species in the first edition.

Page 20

Halecium halecinum. Halecium gracile Verrill, 1874 is regarded as a junior subjective synonym (see Calder 1975).

Halecium scutum. The date was incorrectly cited as 1876 in the first edition (see Calder and Stephens 1997).

Halecium sessile. The date was incorrectly cited as 1866 in the first edition (see Cornelius 1995a).

Halecium undulatum. We add this to the species listed in this edition; it was included as *H. tenellum* (pro parte) in the first edition (see Cornelius 1995a).

Hydrodendron mirabile. Ophiodissa caciniformis (Ritchie, 1907) is regarded as a junior subjective synonym (see Cornelius 1995a).

Sagamihydra dyssymetra. This species was listed as *Endothecium dyssymetrum* in the first edition. *Endothecium* Fraser, 1935 (Hydroida) is an invalid junior homonym of *Endothecium* Koker, 1924 (Anthozoa), and has been replaced by *Sagamihydra* (see Hirohito 1995).

Kirchenpaueriidae. One of the species included in Kirchenpaueriidae here (*Kirchenpaueria plumularioides*) was assigned to the Plumulariidae, as *Plumularia plumularioides*, in the first edition. Another (*Pycnotheca allmani*) was included in the Aglaopheniidae, as *Diplocheilus allmani*, in that edition.

Kirchenpaueria paucinema. We add this to the species listed in this edition (see Fraser 1940).

Kirchenpaueria plumularioides. The date was incorrectly cited as 1876 in the first edition (see Calder and Stephens 1997); it was listed as *Plumularia plumularioides*, family Plumulariidae.

Pycnotheca allmani. This species was listed as *Diplocheilus allmani*, family Aglaopheniidae, in the first edition.

Ventromma halecioides. Fraser (1937) reported this species from the Pacific Northwest, but the record needs verification. *Plumularia inermis* Nutting, 1900 is regarded as a junior subjective synonym (see Calder 1997).

Plumulariidae. One of the species included in this family in the first edition [as *Plumularia plumularioides* (Clark, 1876)], is assigned here to the Kirchenpaueriidae (as *Kirchenpaueria plumularioides*).

Dentitheca dendritica. We add this to the species listed in this edition; specimens have been identified from the coast off Palm Beach, Florida. *Sphaerocystis heteronema* Fraser, 1943 is regarded as a junior subjective synonym (see Calder 1997).

Hippurella annulata. Added to the species listed in this edition, it is now regarded as valid (see Calder 1997).

Monotheca margaretta. *This species was listed as* Plumularia margaretta *in the first edition (see Calder 1997).*

Nemertesia antennina. The date was incorrectly cited as 1767 in the first edition.

Nemertesia irregularis. N. irregularis (Fraser, 1938) is an invalid secondary homonym of *N. irregularis* (Quelch, 1885).

Page 21

Plumularia floridana. Plumularia alicia Torrey, 1902, included as a valid species in the first edition, is regarded as a junior subjective synonym (see Calder 1997).

Plumularia septata. The author and date were mistakenly enclosed in parentheses in the first edition.

Plumularia setacea. Plumularia corrugata Nutting, 1900, included as a valid species in the first edition, is regarded as a junior subjective synonym (see Calder 1997).

Plumularia strictocarpa. This species has sometimes been misidentified as *Plumularia setaceoides* Bale, 1881 in the western North Atlantic; that species was included in the first edition but is excluded here (see Calder 1997).

Halopterididae. One of the species included in Halopterididae in this edition (*Nuditheca dallii*) was assigned to the Aglaopheniidae in the first edition.

Antennella secundaria. The date was incorrectly cited as 1788 in the first edition.

Antennella gracilis Allman, 1877, included as a valid species in the first edition, is regarded as a junior subjective synonym (see Calder 1997; Schuchert 1997).

Halopteris alternata. We add this to the species listed in this edition; it was included in the first edition as *Halopteris diaphana* (Heller, 1868), a species now known only from the Mediterranean Sea and from Brazil (see Schuchert 1997).

Halopteris clarkei. This species was originally described as *Plumularia gracilis* Clarke, 1879, an invalid junior primary homonym of *Plumularia gracilis* de Blainville, 1830, and renamed by Nutting (1900).

Halopteris tenella. This was listed as *Schizotricha tenella* in the first edition (see Schuchert 1997).

Nuditheca dallii. The specific name was misspelled *dalli* and the date was incorrectly cited as 1876 in the first edition (see Calder and Stephens 1997). This species was assigned to the Aglaopheniidae in the first edition but is referred here to the Halopterididae, following Calder (1997).

Aglaopheniidae. One of the species included in this family in the first edition (*Diplocheilus allmani* Torrey, 1902) is assigned here to the Kirchenpaueriidae (as *Pycnotheca allmani*). Another (*Nuditheca dallii*) is transferred to the Halopterididae. *Lytocarpia distans* (Allman, 1877) is removed from the list here because its known bathymetric occurrence is below 200 m.

Aglaophenia apocarpa. Aglaophenia lophocarpa Allman, 1877, *Aglaophenia aperta* Nutting, 1900, *Aglaophenia cristifrons* Nutting, 1900, and *Aglaophenia raridentata* Fraser, 1944 are regarded as junior subjective synonyms (see Bogle 1975).

Aglaophenia dubia. We add this to the species listed in this edition. The western North Atlantic *Aglaophenia dubia* has sometimes been misidentified as *Aglaophenia acacia* Allman, 1883 and *Aglaophenia elongata* Meneghini, 1845 (Calder 1997); those species were included in the first edition. *Aglaophenia flowersi* Nutting, 1900 is regarded as a junior subjective synonym (see Calder 1997).

Aglaophenia latecarinata. Aglaophenia perpusilla Allman, 1877, *Aglaophenia minuta* Fewkes, 1881, *Aglaophenia perforata* Allman, 1885, *Aglaophenia mammillata* Nutting, 1900, and *Aglaophenia minima* Nutting, 1900 are regarded as junior subjective synonyms (see Calder 1997). *Aglaophenia perpusilla* and *A. mammillata* (as *Aglaophenoides mammillatus*) were included as valid species in the first edition. Hydroids identified as *Aglaophenia pelagica* from pelagic *Sargassum* along the Atlantic coast (Fraser 1944) may have been based on this species (Calder 1997).

Page 22

Aglaophenia pluma. The date was incorrectly cited as 1767 in the first edition.

Aglaophenia rhynchocarpa. Aglaophenia cylindrata Versluys, 1899 and *Aglaophenia insolens* Fraser, 1943 are regarded as junior subjective synonyms (see Calder 1997).

Aglaophenia trifida. Aglaophenia rigida Allman, 1877 is regarded as a junior subjective synonym; also, Fraser's reports of *Aglaophenia acacia* from North America are believed to have been based on specimens of *A. trifida* (see Calder 1997). *Aglaophenia dichotoma* Kirchenpauer, 1872 from Tampa Bay is possibly conspecific.

Cladocarpus integer. Cladocarpus pourtalesi Verrill, 1879 is regarded as a junior subjective synonym (see Broch 1918).

Macrorhynchia allmani. Macrorhynchia mercatoris (Leloup, 1937), included as a valid species in the first edition, is regarded as a junior subjective synonym (see Calder 1997).

Macrorhynchia philippina. Aglaophenia tricuspis McCrady, 1859 from South Carolina may have been identical with this species, but it is better considered a *nomen dubium* (Calder 1997). The common name is changed herein.

Lafoeidae. Hebellidae was recognized at the rank of family in the first edition, with species assigned therein to the genera *Hebella, Hebellopsis,* and *Scandia.* The taxon is here recognized as a subfamily of Lafoeidae, and genera referred to it are incorporated below, between those of Lafoeinae (*Acryptolaria, Cryptolaria, Filellum, Grammaria, Lafoea*) and Zygophylacinae (*Zygophylax*). One of the lafoeid species listed in the first edition [*Eucryptolaria pinnata* (Fraser, 1938) (=*Cryptolaria pinnata*)] is deleted because it occurs outside the geographic range covered here. *Hebella calcarata* auct. [not *Laodicea calcarata* A. Agassiz, in L. Agassiz, 1862], reported from various locations along the Atlantic coast (Fraser 1944), is of uncertain identity but is considered likely conspecific with *Hebellopsis scandens* (Calder 1991; Boero, Bouillon, and Kubota 1997).

Grammaria abietina. The date was incorrectly cited as 1851 in the first edition. We add this species to the Pacific coast list in this edition (see Fraser 1937).

Grammaria immersa. This hydroid is similar to and possibly conspecific with *Grammaria gracilis* Stimpson, 1854.

Lafoea dumosa. The date was incorrectly cited as 1828 in the first edition. *Hebella pocillum* is regarded as a junior subjective synonym (see Cornelius 1975).

Lafoea fruticosa. Lafoea symmetrica Bonnevie, 1899, listed as valid in the first edition, is regarded as a junior subjective synonym (see Broch 1918). *Lafoea fruticosa* is frequently considered conspecific with *L. dumosa* (see Cornelius 1995a).

Lafoea gracillima. Regarded as conspecific with *Lafoea fruticosa* and not included in the first edition. The status of *Lafoea dumosa, L. fruticosa,* and *L. gracillima* is still open to review.

Lafoea tenellula. We add this to the species listed in this edition (see Fraser 1944).

Hebellopsis cylindrica. The date was incorrectly cited as 1884 in the first edition. Vervoort (personal communication) considers this a doubtful species and regards records of *H. cylindrica* from North America as based on specimens of *Hebellopsis scandens* (Bale, 1888). *Hebellopsis* Had i, 1913 was regarded as a junior subjective synonym of *Hebella* Allman, 1888 by Boeroet al. (1997).

Scandia corrugata. This was listed as *Hebella corrugata* (Fraser 1938) in the first edition. The unique gonophore morphology of this species sets it apart from others assigned to *Hebella* and *Scandia*, and a new genus may be warranted for it.

Page 23

Zygophylax carolinus. The specific name *carolinus* is considered a noun in apposition, and thus the ending has been modified (Vervoort, personal communication). The author and date were mistakenly not enclosed in parentheses in the first edition.

Zygophylax cervicornis. The author and date were mistakenly not enclosed in parentheses in the first edition.

Zygophylax convallarius. Records of *Zygophylax convallarius* in the Pacific, as listed in the first edition, need verification. Fraser (1944) regarded *Zygophylax cervicornis* from the Pacific as conspecific; this was questioned by Vervoort (1972) and Rees and Vervoort (1987). The specific name *convallarius* is considered a noun in apposition, and thus the ending has been modified (Vervoort, personal communication).

Zygophylax crassitheca. The author and date were mistakenly not enclosed in parentheses in the first edition.

Zygophylax pinnatus. We add this to the species listed in this edition; it is given as *Lafoea pinnata* in Fraser (1944) (see Rees and Vervoort 1987).

Zygophylax reflexus. The author and date were mistakenly not enclosed in parentheses in the first edition.

Zygophylax rigidus. Fraser (1944) referred the extralimital *Euperisiphonia rigida* Fraser, 1940 to the genus *Zygophylax*, but it is not a senior homonym of *Zygophylax rigidus* in this list because it is now assigned to *Cryptolaria* (see Rees and Vervoort 1987).

Zygophylax robustus. This is listed as *Halecium robustum* Verrill, 1873 in Fraser (1944) (see Rees and Vervoort 1987).

Campanulariidae. Two species listed in the first edition (*Campanularia brevicaulis* Nutting, 1915; *Clytia laxa* Fraser, 1937), both from the West Indies, have been removed because they are extralimital to the area of coverage. *Clytia inconspicua* (Forbes, 1848), reported from both the Atlantic and the Pacific coasts of North America (Fraser, 1937, 1944) but omitted in the first edition as a *nomen dubium*, has now been placed on the Official Index of Rejected and Invalid Specific Names in Zoology (ICZN Opinion 1465). *Laomedea angulata* (Hincks, 1861) has been reported from the east coast of North America (Fraser, 1944), but Cornelius (1995b) considered records of the species from areas outside Europe and the Mediterranean as unreliable. *Campanularia pygmaea* Clark, 1875 from Casco Bay, Maine, and included in Fraser (1944), is a *nomen dubium* (Calder and Stephens 1997).

Campanularia emarginata. This species is possibly conspecific with *Campanularia hincksii* (see Calder 1991).

Campanularia volubilis. *Campanularia fusiformis* Clark, 1876 (listed as a valid species in the first edition) and *Campanularia urceolata* Clark, 1877 are regarded as junior subjective synonyms (see Calder and Stephens 1997).

Clytia denticulata. The date was incorrectly cited as 1876 in the first edition (see Calder and Stephens 1997).

Clytia exilis. This species is possibly conspecific with *Clytia gracilis* (see Calder 1991).

Clytia gracilis. We add this to the species listed in this edition (see Calder 1991). *Clytia cylindrica* L. Agassiz, 1862, *Clytia pelagica* van Breemen, 1905, and *Clytia elsaeoswaldae* Stechow, 1914 are regarded as junior subjective synonyms (see Calder 1991).

Clytia hemisphaerica. The date was incorrectly cited as Linnaeus, 1758 in the first edition.

Clytia johnstoni (Alder, 1856), *Clytia raridentata* (Alder, 1861), *Clytia bicophora* L. Agassiz, 1862, *Campanularia gigantea* Hincks, 1866, *Clytia coronata* (Clarke, 1879), *Clytia edwardsi* (Nutting, 1901), and *Clytia minuta* (Nutting, 1901) are regarded as junior subjective synonyms (see Cornelius 1982; Calder 1991). However, this species, and indeed the entire genus *Clytia*, is in need of reappraisal.

Clytia hendersoni. We add this to the species listed in this edition (see Fraser 1937); it was earlier considered conspecific with *Clytia linearis*.

Clytia linearis. *Clytia fragilis* Congdon, 1907 is regarded as a junior subjective synonym (see Calder 1991).

Clytia noliformis. As *Clytia noliformis* McCrady, 1859 in the first edition. McCrady's nominal species is likely synonymous with *C. hemisphaerica*, whereas *C. noliformis* auct. is a distinct species (see Calder 1991; Lindner and Calder 2000).

Clytia paulensis. *Clytia longitheca* (Fraser, 1914) is here considered a probable synonym.

Gonothyraea loveni. We add this species to the Pacific coast list in this edition (see Mills and Miller 1987, *as Gonothyraea clarki*). *Laomedea* (*Gonothyraea*) *clarkii* Marktanner-Turneretscher, 1895 and *Gonothyraea clarki* Torrey, 1902 are considered junior subjective synonyms of *G. loveni*, as well as primary homonyms of each other. A nomenclatural problem will thus arise should *G. clarki* from the west coast be considered a distinct species.

Laomedea neglecta. *Campanularia fragilis* Hincks, 1863 is regarded as a junior subjective synonym (see Cornelius 1995b).

Obelia bidentata. *Obelia bicuspidata* Clark, 1875 and possibly *Clytia longicyatha* (Allman, 1877) are regarded as junior subjective synonyms (see Calder 1991).

Obelia dichotoma. *Obelia hyalina* Clarke, 1879, *Obelia fragilis* Calkins, 1899, *Obelia gracilis* Calkins, 1899, *Obelia griffini* Calkins, 1899, *Obelia surcularis* Calkins, 1899, *Obelia ubia* Nutting, 1901, *Obelia obtusidens* (Jäderholm, 1904), *Obelia equilateralis*

Fraser, 1938, and *Gonothyraea integra* Fraser, 1940 are regarded as junior subjective synonyms (see Calder 1991).

Obelia longissima. Obelia articulata (A. Agassiz, 1865), *Obelia commissuralis* McCrady, 1859, *Obelia flabellata* (Hincks, 1866), *Obelia borealis* Nutting, 1901, and possibly *Obelia corona* Torrey, 1904 and *Campanularia fasciculata* Fraser, 1941, are regarded here as junior subjective synonyms.

Obelia racemosa. We add this to the species listed in this edition (see Fraser 1944); it was regarded as conspecific with *O. dichotoma* in the first edition.

Orthopyxis compressa. The date was incorrectly cited as 1876 in the first edition (see Calder and Stephens 1997).

Page 24

Orthopyxis integra. Orthopyxis caliculata (Hincks, 1853) and *Bonneviella gracilis* Fraser, 1939 are regarded as junior subjective synonyms (see Calder 1970).

Tulpa crenata. This was given as *Campanularia crenata* in the first edition. *Campanularia speciosa* Clark, 1877 and *Campanularia magnifica* Fraser, 1913 are regarded as junior subjective synonyms (see Broch 1918; Calder and Stephens 1997).

Cnidoscyphus marginatus. This was given as *Thyroscyphus marginatus* in the first edition.

Syntheciidae. Two syntheciid species listed in the first edition have been excluded because they are extralimital. The known range of *Synthecium robustum* Nutting, 1904 (not *Synthecium robustum* Totton, 1930) is limited to the Caribbean Sea. *Synthecium marginatum* (Allman, 1877) has been reported off the coast of Florida but at depths below 200 m and is therefore removed from the list

Hincksella cylindrica. Synthecium gracile Fraser, 1937, included as a valid species in the first edition, is regarded as a junior subjective synonym (see Calder 1991).

Hincksella projecta. This was given as *Synthecium projectum* in the first edition.

Synthecium tubithecum. Synthecium nanum Fraser, 1943 is regarded as a junior subjective synonym (see Calder 1991).

Sertulariidae. Sertulariids having hydrothecae arranged in more than two longitudinal series are assigned here to two distinct genera rather than one (i.e., *Selaginopsis* Allman, 1876), as in the first edition. Those resembling *Sertularia* Linnaeus, 1758 in having distinct marginal teeth have been assigned to the genus *Pericladium* Allman, 1876, whereas those allied to *Thuiaria* Fleming 1828 in having an entire hydrothecal margin are retained in *Selaginopsis*. This corresponds with Allman's (1876) original diagnoses of the genera and with the classification recently adopted by Hirohito (1995). Five species of sertulariids listed in the first edition [*Dynamena anceps* (Fraser, 1938), *Dynamena stabilis* (Fraser, 1948), *Symplectoscyphus incisus* (Fraser, 1938), *Symplectoscyphus multinodus* (Fraser, 1948), and *Tridentata simplex* (Fraser, 1938)] have been excluded because their known ranges extend north only as far as Mexico in the Pacific Ocean. Three others [*Dynamena subtilis* (Fraser, 1937), from Puerto Rico, *Sertularella solitaria* Nutting, 1904, from the Bahamas, *Symplectoscyphus sieboldi* (Kirchenpauer, 1884) from Cuba] are extralimital to the Atlantic coasts of Canada and the United States. Finally, *Sertularella magna* Nutting, 1904 from the first edition is excluded here because it is known only from depths exceeding 200 m.

Abietinaria anguina. Naumov (1960) regarded *Abietinaria anguina* and *A. variabilis* as synonyms.

Abietinaria costata. Naumov (1960) included this taxon as a subspecies of *Abietinaria filicula.*

Abietinaria gigantea. The date was incorrectly cited as 1876 in the first edition (see Calder and Stephens 1997).

Abietinaria inconstans. The date was incorrectly cited as 1876 in the first edition (see Calder and Stephens 1997).

Abietinaria kincaidi. This was given as *Diphasia kincaidi* in the first edition (see Naumov 1960).

Abietinaria pulchra. This was given as *Diphasia pulchra* in the first edition (see Naumov 1960).

Abietinaria thuiarioides. The date was incorrectly cited as 1876 in the first edition (see Calder and Stephens 1997).

Abietinaria turgida. The date was incorrectly cited as 1876 in the first edition (see Calder and Stephens 1997).

Abietinaria variabilis. The date was incorrectly cited as 1876 in the first edition (see Calder and Stephens 1997). Naumov (1960) regarded *Abietinaria variabilis* as a synonym of *A. anguina.*

Amphisbetia operculata. We add this to the species listed in this edition (see Mills and Miller 1987, *as Dynamena operculata*).

Diphasia pinastrum Cornelius (1995b) applied the name *D. alata* (Hincks, 1855) to this species, but the binomen *D. pinastrum* (Cuvier, 1830) has priority and avoids further nomenclatural confusion.

Dynamena crisioides. We remove this from the Pacific coast list in this edition; it has been reported only as far north as Mexico (Fraser, 1948).

Dynamena dalmasi. We remove this from the Pacific coast list in this edition; it has been reported only as far north as Mexico (Fraser, 1948).

Page 25

Dynamena disticha. Sertularia cornicina auct., *Sertularia exigua* Allman, 1877, *Sertularia mayeri* Nutting, 1904, and *Sertularia pourtalesi* Mayer, 1904 are regarded as junior subjective synonyms (see Calder 1991).

Dynamena pumila. This species was reported from the Pacific coast (California) by Clark (1876), a record repeated in Fraser (1937), but its occurrence there is doubtful (see Naumov 1960).

Fraseroscyphus sinuosus. We add this to the species listed in this edition (see Boero and Bouillon 1993).

Pericladium mirabilis. This was given as *Selaginopsis mirabilis* in the first edition. Under Allman's (1876) original diagnoses of *Selaginopsis* and *Pericladium*, this species is referable to the latter in having distinct marginal teeth.

Pericladium trilateralis. This was given as *Selaginopsis trilateralis* in the first edition. In having marginal teeth, this species is referable to *Pericladium* rather than *Selaginopsis* (see *Pericladium mirabilis*, above).

Selaginopsis cylindrica. The date was incorrectly cited as 1876 in the first edition (see Calder and Stephens 1997). *Selaginopsis pinnata* Mereschkowsky, 1878 and *Selaginopsis ornata* Nutting, 1904, both included as valid in the first edition, are regarded as junior subjective synonyms (see Naumov 1960).

Sertularella clarkii. The specific name was misspelled *clarki* in the first edition; the correct original spelling is *clarkii*.

Sertularella conella. We add this to the species listed in this edition. Stechow (1920) applied this name to hydroids from the Pacific Northwest that were misidentified as *Sertularella conica* by Calkins (1899) and Fraser (1911).

Sertularella conica. Records of this species from the Pacific Northwest (Fraser, 1937) are considered erroneous (see Calder 1991).

Sertularella diaphana. Sertularella distans (Allman, 1877) [a junior secondary homonym of *Sertularella distans* (Lamouroux, 1816)], *Sertularella pinnata* (Allman, 1877) [a junior secondary homonym of *Sertularella pinnata* (Templeton, 1836)], *Sertularella pinnigera* Hartlaub, 1901, and *Sertularella speciosa* Congdon, 1907 are regarded as junior subjective synonyms (see Calder 1991).

Sertularella flabella. This was listed as *Dictyocladium flabellum* in the first addition. As noted by Vervoort (personal communication), the hydrothecal margin of this species has four cusps instead of three (as in *Dictyocladium*), and it may be referable to *Sertularella* or a new genus. If retained in *Sertularella, S. flabella* (Nutting, 1904) becomes an invalid junior secondary homonym of *S. flabella* (Allman, 1885).

Sertularella ellisi. This name replaces *S. gaudichaudi* (see Medel and Vervoort 1998).

Sertularella tenella. The date was incorrectly cited as 1857 in the first edition. *Sertularella geniculata* Hincks, 1874, included as a valid species in the first edition, is regarded as a junior subjective synonym (see Calder and Vervoort 1998).

Sertularia argentea. We add this to the species listed in this edition; it was considered a synonym of *Sertularia cupressina* in the first edition (see Cornelius 1995b).

Sertularia carolinensis. This species has not been reported since its original description. Fraser (1944) believed it was closely related to *Sertularia argentea.* Based on its description and provenance, *Sertularia plumulifera* (Allman, 1877) may prove conspecific.

Sertularia cupressina. The common name "whiteweed hydroid" was employed in the first edition. The vernacular name "sea cypress" applied to this species by Cornelius (1995b) is adopted here.

Sertularia dalli. The date was incorrectly cited as 1876 in the first edition. The species is added to the Pacific coast list in this edition (see Fraser 1937, as *Thuiaria dalli*).

Sertularia plumosa. The date was incorrectly cited as 1876 in the first edition (see Calder and Stephens 1997).

Sertularia plumulifera. This nominal species may prove synonymous with *Sertularia carolinensis* Verrill, 1872.

Sertularia robusta. The date was incorrectly cited as 1876 in the first edition (see Calder and Stephens 1997). We add this to the Atlantic coast list in this edition (see Fraser 1944, as *Thuiaria robusta*).

Sertularia similis. The date was incorrectly cited as 1876 in the first edition (see Calder and Stephens 1997).

Symplectoscyphus amphoriferus. The species name was *amphorifer* in the first edition; the word should be an adjective, and thus the ending has been modified to match the gender of the genus.

Symplectoscyphus pinnatus. The date was incorrectly cited as 1876 in the first edition (see Calder and Stephens 1997); the species was also listed therein under the binomen *Sertularella pinnata.*

Page 26

Thuiaria alba. This was given as *Salacia alba* in the first edition.

Thuiaria alternitheca. This was given as *Selaginopsis alternitheca* in the first edition (see Naumov 1960).

Thuiaria articulata. This was given as *Salacia articulata* in the first edition (see Cornelius 1995b). *Thuiaria lonchitis* (Ellis and Solander, 1786) and

Thuiaria kolaeensis Jäderholm, 1907 are regarded as junior subjective synonyms.

Thuiaria carica. This was given as *Salacia carica* in the first edition (see Naumov 1960); the species is added to the Pacific coast list in this edition (see Fraser 1937).

Thuiaria desmoides. This was given as *Salacia desmoides* in the first edition.

Thuiaria fraseri. This was given as *Salacia fraseri* in the first edition.

Thuiaria laxa. This was given as *Salacia laxa* in the first edition (see Naumov 1960). *Thuiaria immersa* Nutting, 1904 is considered a junior subjective synonym (see Broch 1918).

Tridentata distans. We add this species to the Pacific coast list in this edition based on Fraser's (1948) report of *Sertularia stookeyi*, a junior subjective synonym, from Southern California. *Sertularia gracilis* Hassall, 1848 is also regarded as a junior subjective synonym (see Calder 1991).

Tridentata marginata. Sertularia amplectens Allman, 1885 and *Sertularia inflata* (Versluys, 1899) are regarded as junior subjective synonyms (see Calder 1991).

Actinulida. As in the first edition, this order of benthic and polyplike hydrozoans is listed here for convenience with the hydroids.

Order Leptothecatae. See page 54.

Melicertum octocostatum. The hydroid stage has been reported from the Atlantic but not the Pacific (Arai and Brinckmann-Voss 1980).

Aequoreidae. Several species listed in the first edition—*Aequorea macrodactyla, A. pensilis, Zygocanna vagans*—have been deleted in this edition because they are outside the reporting range. Characters of several species within the genus are poorly defined, and new records should be evaluated carefully.

Aequorea albida. We add this species to the Pacific coast list in this edition (Bigelow, 1913 as *A. aequorea* var. *albida*).

Aequorea forskalea. We add this species to the Pacific coast list in this edition (Sugisaki, Brodeur, and Napp 1998).

Aequorea victoria. The hydroid was described from the laboratory but not from field-collected specimens.

Blackfordia virginica. We add this species to the Pacific coast list in this edition; it has also been collected in estuaries (Mills and Rees 2000).

Phialella fragilis and *Phialella zappai.* The medusoid stages of these two species were not field collected for the reporting range but were raised in the laboratory from the field-collected hydroids (Boero 1987); however, the field-collected medusa of *Phialella fragilis* was reported from Japan (Uchida 1938).

Page 27

Ptychogena lactea Although the medusa has been reported for A-P-Ac, its connection to its hydroid was deduced from zoogeographical evidence only (Calder 1970; Bouillon 1985).

Tiarannidae. This family was listed in the Anthomedusae in the first edition but is now listed in the Leptothecatae.

Modeeria rotunda. This species occurs outside the continental shelf but within 200 miles of U.S. and Canadian land (Brinckmann-Voss and Arai 1998; Wrobel and Mills 1998).

Cirrholoveniidae. The family Lovenellidae (first edition) is split into Cirrholoveniidae, Eucheilotidae, and Lovenellidae in this edition, following Bouillon (1984). See also Cornelius (1995).

Eucheilotidae. See Cirrholoveniidae.

Eucheilota bakeri. See page 54.

Lovenellidae. See Cirrholoveniidae.

Lovenella gracilis. The medusa was raised from the hydroid. The taxonomic status of this species is discussed in Calder (1971).

Eirenidae. Family Campanulinidae in the first edition has been changed to Eirenidae in this edition (see Bouillon 1985; Cornelius 1995).

Page 28

Campanulariidae. For remarks about the family, see page 57. *Clytia bicophora,* listed in the first edition has been deleted in this edition.

Clytia hemispherica. The name is considered the senior synonym of *C. languida* (Calder 1991).

Obelia dichotoma and *Obelia longissima.* These species have been added to the list for the Pacific coast in this edition. Although distinctions have not been made for field-collected medusae of both species worldwide, the medusae of both species have been raised in the laboratory from their field-collected Pacific hydroids (*Obelia dichotoma* from Race Rocks, Strait of Juan de Fuca, Canada, and *O. longissima* from floats at Sooke Harbour, B.C., Canada, coll. Brinckmann-Voss). The distinctive characteristics of the medusae are described in Stepanjants et al. (1999).

Orthopyxis compressa. The medusa stage has been added to the list in this edition (see Arai and Brinckmann-Voss 1980)

Order Narcomedusae. *Aegina beebei* Bigelow, 1940; *Cunina tenella* Bigelow, 1909; *Pegantha clara* Bigelow, 1909; *P. martagon* Haekel, 1879; and *Solmaris rhodoloma* (Brandt, 1838) were reported from the waters off Baja California and Southern California but without sufficient published location data to determine whether they occur north of Mexican waters and within 329 km of the coast (see Alvariño and Kimbrell 1987; Alvariño 1999). A small Pacific *Solmaris* species with four tentacles, reported as *S. quadrata* in Arai et al. (1993), has not been definitively identified (see Arai et al. 2000).

Aegina citrea. The common name is taken from Robison and Connor (1999).

Cunina frugifera. This species was reported off British Columbia by Arai et al. (1993).

Cunina globosa. This species was collected off California by Widder et al. (1989) and Haddock and Case (1999).

Solmissus marshalli. The common name is taken from Robison and Connor (1999).

Polypodium hydriforme. Raikova (1973, 1988, 1994) has suggested that this species be removed from the Narcomedusae and placed in a new cnidarian class, Polypodiozoa. Although there are similarities between the complex reproduction of *P. hydriforme* and that of other Narcomedusae, they also differ in several ways. The position of *P. hydriforme* outside the class Hydrozoa has also been supported by the difference in DNA from that of other Cnidaria and its similarity to that of the protist phylum Myxozoa (Siddall et al. 1995; Siddall and Whiting 1999). Meanwhile, controversial suggestions that the phylum Myxozoa be added to the phylum Cnidaria (Kent et al. 1994; Siddall et al. 1995) have been criticized by authors such as Hanelt et al. (1996), Lom and Dykova (1997), and Kim et al. (1999). Until all this is resolved by further observations of myxozoan DNA and life cycles, it is probably best to retain *P. hydriforme* in the Narcomedusae, as done by Bouillon (1995).

Pegantha clara Bigelow, 1909 and *P. martagon* Haeckel, 1879. These species have been reported from the waters off Baja California and Southern California but without sufficient published location data to determine whether they occur north of Mexican waters and within 320 km of the Pacific coast (see Alvariño and Kimbrell 1987; Alvariño 1999).

Pegantha laevis. This species was recorded off California by Haddock and Case (1999).

Pegantha martagon. See *P. clara*, above.

Order Trachymedusae. *Geryonia proboscidalis, Halitrephes valdiviae,* and *Sminthea eurygaster* have been reported from the Pacific but with insufficient published data to determine whether they occur north of Mexican waters and within 320 km of the Pacific coast (see Thuesen 1993 and Wrobel and Mills 1998).

Page 29

Haliscera bigelowi, Halitrephes maasi, and *Halitrephes valdiviae.* These species were recorded in the Atlantic region off the Dry Tortugas by Larson et al. (1991). Although *H. valdiviae* was considered a synonym of *H. maasi* by Kramp (1961, 1965) and Bouillon and Barnett (1999), it has been considered a valid species by Larson et al. (1991) and Wrobel and Mills (1998), and tentatively by Gili et al. (1998).

Haliscera conica and *Haliscera racovitzae.* These species were collected off Southern California by Thuesen and Childress (1994).

Halitrephes maasi. See *Haliscera bigelowi,* above.

Halitrephes valdiviae. See *Haliscera bigelowi,* above.

Benthocodon pedunculata Bigelow, 1913. The species was placed in this genus by Larson and Harbison (1990). A second species of *Benthocodon* observed in Monterey Canyon by Larson et al. (1992) has not been definitely identified.

Colobonema sericeum. The common name is taken from Robison and Connor (1999).

Tetrorchis erythrogaster. This species was collected off Southern California by Thuesen and Childress (1994).

Vampyrocrossota childressi. This species was described by Thuesen (1993) from collections off California.

Subclass Siphonophorae. This taxon now generally considered a subclass of the Class Hydrozoa (see Bouillon et al. 1992) and is divided into three orders: Cystonectae, Physonectae, and Calycophorae. It is very likely that most siphonophore species that have been recorded on either the Atlantic or Pacific side of North America will also occur on the other. The exceptions are the benthic species *Angelopsis globosa* and *Dromalia alexandri, Lensia baryi,* and the neritic species *Chelophyes contorta.*

Rhizophysidae. Alvariño (e.g., Alvariño 1991) recorded specimens of *Epibulia ritteriana* Haeckel, 1888 off California. This species is not considered valid, however, for the reasons discussed in Mackie et al. (1987), and was not included in the first edition. The records are considered to be for young, highly contracted specimens of *Rhizophysa eysenhardti.*

Bathyphysa conifera. This species was collected in the region of the Bahamas (Pugh, unpublished data), and almost certainly it will be found to occur on the Atlantic side of North America.

Apolemia uvaria. There are many species of apolemiids yet to be described, and it is doubtful whether all previous records for this particular species are accurate.

Agalmatidae. A single example (?nectophore) of *Moseria convoluta* was identified by Alvariño (1991) off California. However, since this species has otherwise been found only in the Antarctic Ocean, its occurrence off California is considered unlikely, and so it has been excluded from the list.

Agalma elegans. Previously, there were records for this species only from the Atlantic, but Alvariño (1991) and Margulis and Vereshchaka (1994) have recorded it off California.

Frillagalma vityazi. Arai et al. (1993) recorded this species off British Columbia (first Pacific record); and Pugh (1998), who redescribed the species, recorded it from the Dry Tortugas and the Bahamas (first Atlantic records).

Halistemma amphytridis. This species has been collected in the region of the Bahamas (Haddock and Case 1999), and almost certainly, it will be found to occur on the Atlantic side of North America.

Halistemma striata. This species has been collected in the region of the Bahamas (Pugh, unpublished data), and almost certainly, it will be found to occur on the Atlantic side of North America.

Halistemma transliratum. This species has been collected in the region of the Bahamas (Pugh and Youngbluth 1988b), and almost certainly, it will be found to occur on the Atlantic side of North America.

Page 30

Nanomia cara. It is questionable whether there are any true records for this species from the Pacific side of North America. Confusion seems to have arisen when Totton identified specimens from British Columbia for Mackie as *N. cara.* It now appears that they most likely were *N. bijuga,* but the misidentification was followed for many years. Alvariño (1980, 1991) has recorded individual examples of *N. cara* off California, but such records need checking.

Pyrostephidae. We add the family to this edition to accommodate the genus *Bargmannia.*

Bargmannia amoena. Pugh (1999) described this new species from the Dry Tortugas and the Bahamas.

Bargmannia elongata. In the first edition, this species was included in the family Agalmatidae. However, after a revision of the genus, Pugh (1999) returned it to the family Pyrostephidae, where Totton (1965a) had originally placed it. Because of difficulties in distinguishing net-collected nectophores of this species from those of *Bargmannia amoena,* it is doubtful whether all previous records refer to this species.

Bargmannia lata. Mapstone (1998) described this new species from British Columbia.

Craseoa lathetica. This species has been collected in the region of the Bahamas (Pugh and Harbison 1987), and almost certainly, it will be found to occur on the Atlantic side of North America.

Desmophyes annectens. Gasca (1999) recorded this species in the Gulf of Mexico (first Atlantic record).

Desmophyes haematogaster. This species has been collected in the region of the Bahamas (Pugh 1992c), and almost certainly, it will be found to occur on the Atlantic side of North America.

Maresearsia praeclara. Haddock and Case (1999) gave the first Pacific record off California. It has yet to be recorded on the Atlantic side, although it almost certainly occurs there.

Mistoprayina fragosa. This species has been collected in the region of the Bahamas (Pugh and Harbison (1987), and almost certainly, it will be found to occur on the Atlantic side of North America.

Nectadamas diomedeae. In the first edition this species was included under the name *Nectopyramis diomedeae.* Pugh (1992a) reviewed the subfamily Nectopyramidinae and concluded that the species *diomedeae* no longer belonged in the genus *Nectopyramis,* and so he placed it in a new genus, *Nectadamas.* He recorded its presence in the vicinity of the Bahamas, and almost certainly, it will be found to occur on the Atlantic side of North America.

Praya reticulata. Pugh (1992b) recorded this species in the vicinity of the Bahamas, and almost certainly, it will be found to occur on the Atlantic side of North America.

Prayola urinatrix. This species has been collected in the region of the Bahamas (Pugh and Harbison 1987), and almost certainly, it will be found to occur on the Atlantic side of North America.

Rosacea limbata. This species has been collected in the region of the Bahamas (Pugh and Youngbluth 1988a), and almost certainly, it will be found to occur on the Atlantic side of North America.

Rosacea plicata sensu. Totton (1965a) noted, "For the last fifty years the generic name *Rosacea* has been used, not in the sense in which its originators Quoy and Gaimard probably used it, but for a better known species to which Bigelow (1911b) seems to have been the first to apply the name *R. plicata* Quoy and Gaimard, and to give a proper description. In order not to complicate further the very involved nomenclature of the Prayinae I propose that *Rosacea* shall continue in the sense of Bigelow 1911b; and I designate *R. plicata* Quoy and Gaimard sensu Bigelow as its type species." Despite a further discussion of the situation (Pugh and Harbison 1987), Margulis (1994) unfortunately decided to resurrect *R. plicata* Quoy and Gaimard 1827 and to rename Bigelow's material. There are several errors in her arguments, but her proposition is rejected primarily on the basis that it destabilizes the nomenclature, although the matter may need to be referred to the ICZN.

Rosacea repanda. This species has been collected in the region of the Bahamas (Pugh and Youngbluth

1988a), and almost certainly, it will be found to occur on the Atlantic side of North America.

Vogtia glabra. Although this species has been found in the subtropical North Pacific, it does not appear to have been recorded off the west coast of North America, as was reported in the first edition.

Vogtia serrata. Alvariño (1980) also includes records for *Vogtia kuruae* Alvariño, 1967, but the latter is considered a junior synonym of *V. serrata.*

Page 31

Gilia reticulata. This species was listed as *Lensia reticulata* in the first edition. However, the resemblance of the eudoxid bract to those of certain clausophyid species led Pugh and Pagès (1995) to remove it into a new diphyid subfamily, Giliinae, with a new generic name.

Lensia baryi. It is uncertain whether this species is distinct from *Lensia achilles.* Totton (1965b) distinguished it mainly on the shape of the somatocyst, but the variability of this structure needs to be studied further. Another species, *Lensia cordata* Totton, 1965, which Totton distinguished from *L. achilles* on the same basis, was recorded in the Californian region by Margulis and Vereshchaka (1994). Since Totton's records are mainly from the equatorial region of East Africa, this record is considered dubious and has not been included.

Lensia cossack. Margulis and Vereshchaka (1994) recorded this species off California (first Pacific record).

Lensia exeter. This species was not included in the first edition, but Alvariño (1985) recorded it off Baja California, and almost certainly, it will be found to occur on both the Pacific and the Atlantic sides of North America.

Lensia subtilis. Alvariño (1985) and Margulis and Vereshchaka (1994) have recorded this species off California (first Pacific records).

Lensia subtiloides. In the first edition an Atlantic record, from Margulis (1978) of this species was included, but the location at which it was collected is outside the area of interest and, furthermore, this species is unlikely to occur off the Atlantic coast of North America. The species is largely confined to the western Pacific and Indian Oceans and is generally neritic. Thus the Pacific records (Alvariño 1967, 1980) may be incorrect. However, as Gasca and Suárez (1992) have also recorded it as a rarity off Baja California, the Pacific records are tentatively retained.

Sulculeolaria turgida. Gasca (1999) recorded this species from the Gulf of Mexico, which is the first Atlantic record.

Chuniphyes moserae. Margulis and Vereshchaka (1994) recorded this species off California (first Pacific record).

Clausophyes moserae. Margulis (1988) pointed out that the specimens that Moser (1925) described under the name *Clausophyes ovata* Keferstein and Ehlers, 1860 belonged to a different species. This error was subsequently followed by other researchers. She thus gave Moser's species the new name, *C. moserae.* The true *C. ovata* is quite rare, and it is assumed that all previous North American records for it actually refer to *C. moserae.*

Clausophyes ovata. Margulis and Vereshchaka (1994) recorded this species off California (first Pacific record). It has also been collected in the region of the Bahamas (Pugh, unpublished data), and almost certainly, it will be found to occur on the Atlantic side of North America.

Crystallophyes amygdalina. Margulis (1988) moved this species into the genus *Chuniphyes,* but this is not accepted as valid.

Sphaeronectes irregularis. Margulis and Vereshchaka (1994) recorded it off California (first Pacific record).

Abyla bicarinata. Alvariño (1991) recorded a specimen of *Abyla brownia* Sears, 1953 from California. However, in agreement with Totton (1965a), this species is considered a junior synonym of *A. bicarinata.*

Abyla haeckeli. R. Gasca (1999) recorded this species from the Gulf of Mexico (first Atlantic record).

Page 32

Ceratocymba leuckarti. Gasca and Suárez (1991) recorded this species from the Gulf of California, and it is likely that it also occurs off the Pacific coast of North America.

Enneagonum hyalinum. Alvariño (1991) recorded this species from Baja California, and Gasca and Suárez (1991) recorded it in the Gulf of California, so it is likely that it also occurs off the Pacific coast of North America.

Antipathidae. *Antipathes alopercoides,* listed in the first edition, is omitted in this edition because the taxonomic validity of this species is highly doubtful. It was originally described off North Carolina.

Antipathes atlantica. The common name was introduced by Humann (1993).

Antipathes barbadensis. The common name was introduced by Humann (1993).

Antipathes caribbeana. This species was described after publication of the first edition.

Antipathes expansa. This species was described after publication of the first edition.

Antipathes gracilis. A. brooki, included in the first edition, is now considered a junior synonym of this species. The common name was introduced by Humann (1993).

Antipathes hirta. This species was reported off Florida by Humann (1993).

Antipathes lenta. The common name was introduced by Humann (1993).

Antipathes pennacea. The common name was introduced by Humann (1993).

Antipathes tanacetum. The common name was introduced by Humann (1993).

Antipathes thamnea. This species was only recently discovered off Florida.

Leiopathidae. Opresko (1998) resurrected and emended the diagnosis of the family Leiopathidae Haeckel, 1896.

Leiopathes glaberrima. See Leiopathidae.

Ceriantheomorphe brasiliensis. This species was recorded from Texas by Carlgren and Hedgpeth (1952).

Ceriantheopsis americana. The specific name, misspelled in the first edition, is corrected to *americana.* Although the spelling *americanus* is commonly used, the *-opsis* ending of the generic name requires a feminine ending to the specific name.

Isodactylactis borealis. We add this to the species listed in this edition.

Pachycerianthus fimbriatus. P. torreyi, used by several California authors in the last decade, is a junior synonym of *P. fimbriatus* (see Arai 1971).

Synarachnactis brachiolata. This species was described by A. Agassiz (1863). However, there was a reference to the name in the Proceedings of the Boston Society of Natural History (Agassiz 1862), and the 1862 date for the species is accepted by Leloup (1962). The species was placed in the new genus *Synarachnactis* by Carlgren (1924). It was recorded from the Arctic by Verrill (1922). Although this may be the larva of *Cerianthus borealis,* that has not been proven.

Page 33

Carijoa riisei. Humann (1993) refers to this species as the "white telesto."

Cryptophyton goddarti. This species was described from the Oregon coast.

Telesto fruticulosa. Humann (1993) refers to this species as the "orange telesto."

Alcyonium rudyi. Verseveldt and van Ofwegen (1992) described this species from intertidal waters off Oregon.

Gersemia rubiformis. An anonymous reviewer reports that this species was also collected in the Alaskan Beaufort Sea (Ac).

Siphonogorgia agassizii. An anonymous reviewer reports that this species was collected "along the Pinnacle trend area off Mississippi at depths of less than 200 m."

Keroeididae. We add the family and species in this edition.

Thelogorgia studeri. See Keroeididae.

Briareum asbestinum. Humann (1993) reports that this species is also commonly known as "deadman's fingers."

Diodogorgia nodulifera. Humann (1993) refers to this species as the "colorful sea rod."

Erythropodium caribaeorum. Humann (1993) refers to this species as the "encrusting gorgonian."

Titanideum frauenfeldii. Humann (1993) refers to this species as the "brilliant sea fingers."

Paragorgia arborea. The species was found in the Gulf of Alaska and in the Monterey Canyon (G. Williams, personal communication). The common name "bubble gum coral" was listed in Breeze et al. (1997), along with "rubber trees" and "strawberry plants."

Bebryce grandis. An anonymous reviewer reports that this species was collected "along the Pinnacle trend area off Mississippi at depths of less than 200 m."

Bebryce parastellata. An anonymous reviewer reports this species was collected "along the Pinnacle trend area off Mississippi...at depths of less than 200 m."

Eunicea calyculata. Humann (1993) refers to this species as the "warty sea rod."

Eunicea fusca. Humann (1993) refers to this species as the "doughnut sea rod."

Page 34

Eunicea mammosa. Humann (1993) refers to this species as the "swollen-knob candelabrum."

Eunicea succinea. Humann (1993) refers to this species as the "shelf-knob sea rod."

Hypnogorgia pendula. An anonymous reviewer reports that this species was collected "along the Pinnacle trend area off Mississippi...at depths of less than 200 m."

Lytreia plana. According to Bayer's (1981) scheme of classification of the Octocorallia, this species is now included among the Plexauridae. It was included in the Paramuriceidae in the first edition.

Muricea elongata. Humann (1993) refers to this species as the "orange spiny sea rod."

Muricea laxa. Humann (1993) refers to this species as the "delicate spiny sea rod."

Muricea muricata. Humann (1993) refers to this species as the "spiny sea fan."

Muricea pendula. Humann (1993) refers to this species as the "pinnate spiny sea fan."

Muriceides hirtus. According to Bayer's (1981) scheme of classification of the Octocorallia, this species is now included among the Plexauridae. It was included in the Paramuriceidae in the first edition.

Muriceopsis flavida. Humann (1993) refers to this species as the "rough sea plume."

Paramuricea grandis. According to Bayer's (1981) scheme of classification of the Octocorallia, this species is now included among the Plexauridae. It was included in the Paramuriceidae in the first edition.

Paramuricea placomus. According to Bayer's (1981) scheme of classification of the Octocorallia, this species is now included among the Plexauridae. It was included in the Paramuriceidae in the first edition.

Placogorgia mirabilis. According to Bayer's (1981) scheme of classification of the Octocorallia, this species is now included among the Plexauridae. It was included in the Paramuriceidae in the first edition.

Placogorgia tenuis. According to Bayer's (1981) scheme of classification of the Octocorallia, this species is now included among the Plexauridae. It was included in the Paramuriceidae in the first edition.

Plexaura flexuosa. Humann (1993) refers to this species as the "bent sea rod."

Plexaurella dichotoma. Humann (1993) refers to an unidentified species of this genus as the "slit-pore sea rod."

Plexaurella nutans. Humann (1993) refers to this species as the "giant slit-pore sea rod."

Pseudoplexaura flagellosa. Humann (1993) refers to an unidentified species of this genus as the "porous sea rod."

Scleracis guadalupensis. According to Bayer's (1981) scheme of classification of the Octocorallia, this species is now included among the Plexauridae. It was included in the Paramuriceidae in the first edition.

Scleracis pumila. According to Bayer's (1981) scheme of classification of the Octocorallia, this species is now included among the Plexauridae. It was included in the Paramuriceidae in the first edition.

Spinimuricea atlantica. According to Bayer's (1981) scheme of classification of the Octocorallia, this species is now included among the Plexauridae. It was included in the Paramuriceidae in the first edition. Transferred to this genus by Grasshoff (1992).

Swiftia casta. According to Bayer's (1981) scheme of classification of the Octocorallia, this species is now included among the Plexauridae. It was included in the Paramuriceidae in the first edition.

Swiftia exserta. Humann (1993) refers to this species as the "red polyp octocoral"; he reported it as common in moderate to deep patch reefs and rocky-sandy substrates off Florida. According to Bayer's (1981) scheme of classification of the Octocorallia, this species is now included among the Plexauridae. It was included in the Paramuriceidae in the first edition.

Swiftia kofoidi. According to Bayer's (1981) scheme of classification of the Octocorallia, this species is now included among the Plexauridae. It was included in the Paramuriceidae in the first edition.

Swiftia spauldingi. This species is transferred from *Psammogorgia* to *Swiftia* in this edition.

Thesea nivea. Humann (1993) refers to this species as the "white eye sea spray"; he reported it in depths within safe diving limits off the north Florida Atlantic coast. Marques-Paraense (1996) stated that this species does not belong to the genus *Thesea.* According to her, the outer coenenchymal sclerites from the holotype (MCZ 4645) present outer and inner surfaces homogeneous. The erroneous combination is here maintained because Marques-Paraense did not indicate to which genus it should be assigned.

Villogorgia nigrescens. According to Bayer's (1981) scheme of classification of the Octocorallia, this species is now included among the Plexauridae. It was included in the Paramuriceidae in the first edition.

Page 35

Leptogorgia cardinalis. Grasshoff (1988) established the equivalence of the genera *Lophogorgia* and *Leptogorgia*, the latter being the senior synonym.

Leptogorgia caryi. Euplexaura marki Kükenthal, 1913 of the first edition is now considered a junior synonym of this species.

Leptogorgia chilensis. Satterlie and Case (1979) worked with *L. chilensis* from California in a paper on physiology of gorgonians. Also, see remarks for *L. cardinalis.*

Leptogorgia hebes. See remarks for *L. cardinalis.*

Leptogorgia miniata. See remarks for *L. cardinalis.*

Leptogorgia punicea. See remarks for *L. cardinalis.*

Leptogorgia virgulata. Humann (1993) refers to this species as the "colorful sea whip."

Pseudopterogorgia bipinnata. Humann (1993) refers to this species as the "bipinnate sea plume."

Pseudopterogorgia elisabethae. Humann (1993) refers to species of this genus in general as "sea plumes."

Pterogorgia guadalupensis. Humann (1993) refers to this species as the "grooved-blade sea whip."

Calcaxonia. This new suborder was established by Grasshoff (1999).

Ellisella schmitti. Humann (1993) and Bayer and Grasshoff (1995) reported this species for the study area. Humann (1993) refers to this species as the "bushy sea whip."

Nicella goreaui. Humann (1993) refers to this species as the "orange deep water fan"; he reported its occasional occurrence off southern Florida, rare within safe diving limits.

Riisea paniculata. Bayer and Grasshoff (1994) reported this species from the "Straits of Florida," without indication of depth. It has been reported from other localities in depths of 136–471 m (Deichmann 1936). It is tentatively included in the North American list.

Arthrogorgia utinomii. This species was described from the Aleutians (163–234 m) by Bayer (1996).

Fanellia compressa. We add this to the species listed in this edition (see Bayer 1982).

Fanellia fraseri. This species was reported from the Gulf of Alaska (82–183 m). Possibly, it is identical with *F. compressa* (Verrill, 1865), according to Kükenthal (1924). It has been transferred from genus *Callogorgia* to *Fanellia* (see Bayer 1982).

Primnoa resedaeformis. The common name "red trees" was listed by Breezeet al. (1997), as well as "sea corn."

Page 36

Virgulariidae. Williams (1995) considered three species (*Ptilosarcus quadrangularis* Moroff, 1902; *P. sinuosus* Gray, 1860; and *P. verrillii* Pfeffer, 1886) invalid. Therefore, they are removed from the North American list.

Acanthoptilum agassizi. Bayer (1957) recorded this species off Pensacola, Florida, at a depth of 100 fathoms (183 m).

Acanthoptilum oligacis. Bayer (1957) described this species from the Mississippi Delta, at a depth of 100 fathoms (183 m).

Halipteris christii. Williams (1995) synonymized *Halipteris* Kölliker, 1869 and *Balticina* Gray, 1870. Therefore, the species is combined with *Halipteris* in this edition of the North American list.

Halipteris finmarchica. Williams (1995) synonymized *Halipteris* Kölliker, 1869 and *Balticina* Gray, 1870. Therefore, the species is combined with *Halipteris* in this edition of the North American list.

Stylatula antillarum. This species was recorded by Bayer (1957) off Cape San Blas, Florida, and off the Mississippi Delta, at depths shallower than 200 m.

Stylatula elongata. The authorship for this species was erroneously credited to Verrill, 1864 in the first list. It was actually described by Gabb (1862).

Ptilosarcus gurneyi. The species was originally described by Gray (1960) as *Sarcoptilus gurneyi*. Parentheses were mistakenly omitted from the first edition.

Ptilosarcus undulatus. The species was described by Verrill (1865) from Pinnacati Bay, California, as *Leioptilum undulatum*.

Page 37

Anthopleura sola. This is added to the species list in this edition.

Bunodosoma granuliferum. The author's name is corrected to correspond to that in the original publication.

Urticina mcpeaki. This species was described after publication of the first edition.

Urticina piscivora. The date of publication has been corrected.

Actinoscyphia saginata. The common name has been added.

Sicyonis obesa. The author and date have been corrected.

Stomphia pacifica. This species was described after publication of the first edition.

Aiptasia californica. This species was mistakenly omitted from the first edition.

Bartholomea annulata. The common name is changed to correspond to that proposed by Humann (1992).

Alicia mirabilis. A common name is added, as suggested by Humann (1992).

Lebrunia coralligens. A common name is added, as suggested by Humann (1992).

Lebrunia danae. A common name is added, as suggested by Humann (1992).

Page 38

Viatrix globulifera. A common name is added, as suggested by Humann (1992).

Diadumene franciscana. This species was described in an issue of the *Wasmann Journal* bearing the date Fall 1955, but the author (Hand) stamped his reprints "Pub. 14 Feb. 1956," which is thus considered the correct date of publication.

Diadumene lighti. See *D. franciscana*, above, concerning the year of publication.

Diadumene lineata. The date of publication has been changed to 1870.

Drillactis pallida. The author and date have been changed.

Edwardsia elegans. Parentheses are removed from the author's name, and a common name is added, as suggested by Humann (1992).

Edwardsia leidyi. A common name is proposed herein.

Edwardsia sulcata. Parentheses are removed from author's name.

Nematostella vectensis. A common name is proposed herein.

Cactosoma arenaria. The spelling of the species name has been corrected.

Halcampa decemtentaculata. This species was described in the *Wasmann Journal* bearing the date of Fall 1954, but the author (Hand) stamped his reprints "Pub. 7 Jan. 1955," which is thus considered the correct date of publication.

Halcampoides purpureus. The date of publication has been corrected.

Pentactinia californica. This species is added to the Pacific coast list in this edition.

Anemonactis mazeli. This species is added to the Pacific coast list in this edition.

Peachia parasitica. A common name is proposed herein.

Page 39

Hormathia digitata. This species is added to the Pacific coast list in this edition.

Liponema brevicornis. The spelling of the specific name has been corrected.

Liponema multicornis. The spelling of the specific name has been corrected.

Metridium exilis. This species was described in an issue of the *Wasmann Journal* bearing the date Fall 1955, but the author (Hand) stamped his reprints "Pub. 14 Feb. 1956," which is thus considered the correct date of publication. The spelling of the specific name is corrected.

Metridium farcimen. This name has been determined to be the senior synonym of *M. giganteum*, as listed in the first edition.

Metridium senile. The date of publication has been corrected.

Nemanthidae. The addition of *Nemanthus californicus* requires adding this family to North American fauna.

Nemanthus californicus. This species was mistakenly omitted from the first edition.

Epicystis crucifera. The spelling of the generic name is corrected and the common name is changed to be consistent with that proposed by Humann (1992).

Actinothoe californica. This species was mistakenly omitted from the first edition.

Actinothoe modesta. The date of publication has been corrected.

Actinothoe pustulata. The date of publication has been corrected.

Sagartia catalienensis. This species was mistakenly omitted from the first edition.

Sagartiogeton verrilli. The spelling of the generic name has been corrected.

Stichodactyla helianthus. The date of publication has been corrected and the common name amended.

Zoanthus pulchellus. The common name has been changed to be consistent with that of Humann (1992).

Page 40

Epizoanthus americanus. A common name is proposed herein.

Epizoanthus paguriphilus. A common name is proposed herein.

Discosoma carlgreni. A common name is added, as suggested by Humann (1992).

Discosoma neglecta. A common name is added, as proposed by Humann (1992).

Discosoma sanctithomae. A common name is added, as proposed by Humann (1992).

Ricordia floridana. A common name is added, as proposed by Humann (1992).

Madracis mirabilis sensu Wells, 1973. As explained in the first edition, *Madracis mirabilis* (Duchassaing and Michelotti, 1860) is a junior synonym of *M. myriaster* (Milne Edwards and Haime, 1849); however, specimens usually identified as *M. mirabilis* (e.g., by Wells, 1973, and many others) are not *M. myriaster* but an as-yet-undescribed species (see Cairns 1979).

Madracis pharensis forma *pharensis.* Reported as simply *M. pharensis* in the first edition, this species is herein listed in its two forms: the azooxanthellate form *pharensis* and the zooxanthellate form *luciphila.* The nominate form occurs in the Pacific but not the eastern Pacific, off North America (Cairns and Zibrowius 1997). Humann (1993) suggested common names for both forms.

Madracis pharensis forma *luciphila.* See *M. pharensis* forma *pharensis,* above.

Page 41

Colpophyllia natans. Colpophyllia breviserialis and *C. amaranthus,* listed in the first edition, are now considered junior synonyms of *C. natans.*

Favia fragum. The date of authorship has been changed.

Montastraea cavernosa. Parentheses have been added to the author's name.

Montastraea faveolata. This species was previously lumped under *M. annularis,* but Weil and Knowlton (1994) have shown it to be a separate sympatric sibling species, based on ecology, morphology, and molecular evidence.

Montastraea franski. See *M. faveolata,* above.

Solenastrea bournoni. The date of publication has been corrected.

Rhizangiidae. Reanalysis of the records of *Astrangia rathbuni* Vaughan, 1906, a rhizangiid listed in the first edition, indicates that it does not occur north of the latitude of Surinam (Cairns 2000), and it is therefore deleted from the list of North American species. The genera *Colangia* and *Phyllangia* are transferred from this family to the Caryophylliidae, as explained below.

Astrangia haimei. Cairns (1994) considered this species the senior synonym of *A. lajollaensis* Durham, 1947; the latter species is thus deleted from the list.

Astrangia solitaria. Humann (1993) refers to this species as the "dwarf cup coral."

Madrepora carolina. Messing (1987) has called this species the "Pourtalès fan coral."

Madrepora oculata. The western Atlantic depth range of *Madrepora oculata*, listed in the first edition from the Atlantic coast, was revised downward to 300–1,391 m (Cairns 2000), and thus this range is removed from this edition as being too deep. But the species is now known from the Pacific coast of North America in water as shallow as 84 m, and thus the species remains on the list but as a Pacific record (Cairns 1994).

Oculina profunda. This new species was described from California at 119–742 m by Cairns (1991a).

Oculina varicosa. A common name is suggested herein.

Page 42

Dichocoenia stokesi. D. stellaris, listed in the first edition, is now considered a junior synonym of this species.

Isophyllia sinuosa. I. multiflora, listed in the first edition, is considered a junior synonym of *I. sinuosa* (see Cairns 1982).

Mycetophyllia lamarckiana. M. danaana, listed in the first edition, is now considered a junior synonym of this species.

Caryophylliidae. Two caryophylliids—*Tethocyathus cylindraceus* and *Thalamophyllia gombergi*—are removed from the list based on a reevaluation of their depth ranges by Cairns (2000).

Anomocora marchadi. Cairns (2000) recently transferred this species from *Asterosmilia* to *Anomocora.* It occurs in the western Pacific (Cairns and Zibrowius 1997) but is not known from the eastern Pacific off North America.

Anomocora prolifera. Cairns (2000) recently transferred this species from *Asterosmilia* to *Anomocora.*

Caryophyllia alaskensis. Vaughan originally described this species in the genus *Caryophyllia*, and thus the parentheses should be removed from his name.

Caryophyllia barbadensis. Cairns et al. (1994) first reported this species from North America off Louisiana at 129–144 m.

Cladocora arbuscula. The genus was placed in the family Faviidae in the first edition, but Cairns (2000) now considers it a caryophylliid. The date of authorship has been changed.

Cladocora debilis. See *Cladocora arbuscula.*

Coenocyathus humanni. This species was recently described from Florida at 20 m (Cairns 2000). Humann (1993) illustrated the holotypic specimen and provided the common name.

Coenocyathus parvulus. Rezak et al. (1985) and Cairns et al. (1994) reported this species from the northern Gulf of Mexico as *Caryophyllia parvula.* Cairns (2000) later transferred the species to the genus *Coenocyathus.*

Colangia immersa. In the first edition this genus was placed in the family Rhizangiidae, but it is now considered a caryophylliid (Cairns 2000).

Crispatotrochus foxi. This species was listed in the genus *Cyathoceras* in the first edition, but Cairns (1994) transferred it to *Crispatotrochus.*

Desmophyllum dianthus. This name was shown by Cairns (1994) to be the senior synonym of *D. cristagalli*, a species listed in the first edition. Messing (1987) supplied the common name.

Labyrinthocyathus quaylei. This species was listed in the genus *Cyathoceras* in the first edition, but Cairns (1994) transferred it to *Labyrinthocyathus.*

Lophelia pertusa. This species was reported as *Lophelia prolifera* in the first edition. The correct senior synonym of this species described by Linnaeus is *L. pertusa* (see Zibrowius 1980). The common name "spider hazard" was given by Breeze et al. (1997), for its tendency to foul deep-sea fishing nets off Nova Scotia.

Oxysmilia rotundifolia. The correct date of publication is 1848, not 1849 as listed in the first edition.

Paracyathus stearnsii. This name was shown by Cairns (1994) to be the senior synonym of *P. calthus*, which had been listed in the first edition.

Phacelocyathus flos. The common name was suggested by Humann (1983).

Page 43

Phyllangia americana americana. In the first edition this genus was placed in the family Rhizangiidae, but is now considered a caryophylliid (Cairns 2000). Also, the western Atlantic populations of this species are now considered the nominate subspecies, and the eastern Atlantic populations are considered *Phyllangia americana mouchezii* (Lacaze-Duthiers, 1897) (see Cairns 2000).

Phyllangia pequegnatae. Previously reported as *Coenocyathus* n. sp. by Rezak et al. (1985) and *Phyllangia americana* by Cairns et al. (1994), this species was finally described by Cairns (2000). It occurs off Alabama and South Carolina at depths of 48–112 m.

Polycyathus senegalensis. Hubbard and Wells (1986) and Cairns et al. (1994) reported this species from eastern Florida and the northern Gulf of Mexico. It was later redescribed and figured by Cairns (2000).

Premocyathus cornuformis. This species was listed in the genus *Caryophyllia* in the first edition, but Cairns (2000) transferred it to *Premocyathus.*

Thalamophyllia riisei. Viada and Cairns (1987) reported this species from the Gulf of Mexico off Louisiana at 155 m. Humann (1993) supplied the common name.

Turbinoliidae. Considered a subfamily of the Caryophylliidae at the writing of the first edition, this name has been elevated to familial rank (Cairns 1997).

Deltocyathoides stimpsonii. Listed in the genus *Peponocyathus* in the first edition, this species was transferred to *Deltocyathoides* by Cairns (1997).

Sphenotrochus andrewianus moorei. Previously reported as *Sphenotrochus* sp. by Cairns (1978) and Cairns et al. (1994), the subspecies was described as new by Cairns (2000). This tiny coral occurs from North Carolina to Florida at depths of 9–42 m.

Flabellum floridanum. Reported as *Flabellum fragile* in the first edition, this species was renamed *F. floridanum* by Cairns (1991b) to avoid homonymy with the fossil species *Flabellum cuneiforme* var. *fragile* Vaughan, 1900.

Javania cailleti. This species was reported for the first time off western North America (British Columbia) by Cairns (1994).

Javania californica. This species was described as new from Monterey Bay and Cordell Bank, California (62–170 m), in a revision of the Scleractinia of the temperate North Pacific (Cairns 1994).

Polymyces montereyensis. Polymyces tannerensis, listed in the first edition, is now considered a junior synonym of this species (Cairns 1994).

Guynia annulata. This tiny, cryptic species occurs in the Pacific, but it is not yet known from the eastern Pacific off North America (Cairns and Zibrowius 1997).

Stenocyathus vermiformis. This elongate, vermiform species occurs in the Pacific, but it is not known from the eastern Pacific off North America (Cairns 1995).

Gardinariidae. This family was established in 1996 by Stolarski. The nominate genus *Gardineria* had previously been included in the family Flabellidae.

Gardineria paradoxa. This species occurs in the Pacific, but it is not known from the eastern Pacific off North America (Cairns and Zibrowius 1997). In the first edition this species was classified in the family Flabellidae.

Dendrophylliidae. *Dendrophyllia californica* is removed from the list of dendrophylliids because its occurrence, limited to the coast of Baja California, is extralimital.

Balanophyllia palifera. This species was first reported from shallow water off North America (Louisiana, 175 m) by Viada and Cairns (1987).

Cladopsammia manuelensis. Listed in the genus *Rhizopsammia* in the first edition, this species is now considered a *Cladopsammia* (see Cairns 2000).

Dendrophyllia oldroydae. The spelling, authorship, and date have been corrected from the listing in the first edition (see Cairns 1994).

Eguchipsammia cornucopia. Listed in the genus *Dendrophyllia* in the first edition, this species is now considered an *Eguchipsammia* (Cairns 1994).

Eguchipsammia gaditana. Listed in the genus *Dendrophyllia* in the first edition, this species is now considered an *Eguchipsammia* (Cairns 1994). It is known from the Pacific but not from the eastern Pacific off North America.

Eguchipsammia strigosa. This is a species described as new since publication of the first edition; it is known from the North Carolina and Venezuela coasts at depths of 25–77 m (Cairns 2000).

Rhizopsammia goesi. This species was first reported off North America (Mississippi River Delta, off Louisiana) by Cairns (2000).

Tubastraea coccinea. This species is known from Baja California (Cairns 1994) and the northern Gulf of Mexico (S. Viada, personal communication). Cairns (2000) considers it an introduced species.

Aulacoctena acuminata. The family placement was made by Harbison (1996). This species has been seen by Haddock (date?) and others in both the Atlantic and the Pacific.

Haeckelia beehleri. This species was reported from California, near the surface (Haddock and Case 1995, 1999).

Haeckelia bimaculata. This species was reported from California, near the surface (Wrobel and Mills 1998; Haddock and Case 1999); it was also seen by Haddock (date?) in Atlantic waters.

Haeckelia rubra. The correct authorship is Kölliker (1853).

Page 44

Bathyctena chuni. This species is known from California, usually at 1,000–3,500 m (Wrobel and Mills 1998) but is often collected in nonclosing nets at unknown depths; thus it is included in the list in this edition.

Lampea pancerina. This species was reported from California, near the surface (Wrobel and Mills 1998).

Hormiphora californensis. This species was reported from California, from the surface to 200 m (Haddock and Case 1995; Wrobel and Mills 1998).

Pleurobrachia bachei. The author is A. Agassiz (in L. Agassiz 1860).

Euplokamididae. This family has been moved farther down the phylogenetic list of cydippid families, as per Harbison (1996), because of its unusual, highly modified tentacles.

Euplokamis dunlapae. Frank and Widder (1997) report vertical migration by *Euplokamis* species in the Gulf of Maine. C. E. Mills (1993, unpublished) has examined submersible-collected specimens from the same part of the Gulf of Maine and determined that they are *Euplokamis dunlapae.*

Callianira bialata. The author and date are earlier than noted in the first edition.

Charistephane fugiens. This midwater species was collected off California (Wrobel and Mills 1998; Haddock and Case 1999) and seen by Harbison (date?) in the Atlantic.

Undescribed species of mertensiid (with pink-red tentacles). Listed as *Mertensia* species in the first edition and still undescribed, this species has been collected from British Columbia to California (Wrobel and Mills 1998).

Undescribed species "*Agmayeria tortugensis.*" This enormous red cydippid is well-known to submersible users as the Tortugas Red, named for where it was first observed, but it has been reported from several locations on the Atlantic coast of North America. It has not yet been described. The *nomen nudum* "*Agmayeria tortugensis*" was used by Bailey et al. (1995). This ctenophore has been tentatively assigned to the family Mertensiidae (G. R. Harbison, personal communication).

Coeloplana species. As noted in the first edition, this may have been *Vallicula multiformis* (note the correct spelling) (G. R. Harbison, personal communication).

Vallicula multiformis. This species was collected by C. Gramlich (personal communication) and identified by G. I. Matsumoto in San Diego in the warm El Niño year 1997; it was present for only a few weeks before the water cooled and it disappeared.

Thalassocalyce inconstans. This species was reported from California, in surface to midwater (Wrobel and Mills 1998; Haddock and Case 1999).

Bolinopsis infundibulum. This species was reported from Point Barrow, Alaska, in the Arctic by MacGinitie (1955).

Undescribed species "*Lampoctena sanguineventer.*" This spectacular deep-red lobate has been seen in both Pacific and Atlantic waters off North America (G. R. Harbison and G. I. Matsumoto, personal communication). It has yet not been described, but the *nomen nudum* "*Lampoctena sanguineventer*" was used by Nybakken (1993 and 1997) and has also been used on the World Wide Web prior to 2000. A formal description, placing this species in a new family, is in preparation (G. R. Harbison and G. I. Matsumoto, personal communication).

Page 45

Eurhamphaea vexilligera. This species was reported from the Pacific coast (Southern California, location unpublished) by Wrobel and Mills (1998).

Deiopea kaloktenota. This species was mistakenly included in the family Bolinopsidae in the first edition. Harbison (1996) puts it in Eurhampheidae; C. E. Mills (1998–2000) puts it in its own family Deiopeidae. It was reported from California (Wrobel and Mills 1998; Haddock and Case 1999).

Kiyohimea aurita. This species is known from the Atlantic and Pacific coasts from submersible observations (Haddock and Mills, unpublished).

Kiyohimea usagi. A new species, this was described from California by Matsumoto and Robison (1992).

Leucothea pulchra. This species seems to be a Pacific species; we believe that the Atlantic distribution noted in the first edition was in error.

Ocyropsis crystallina crystallina. The date of publication is 1828, not 1826, as in the first edition.

Ocyropsis maculata maculata. The date of publication is 1828, not 1826, as in the first edition. This species has been reported from Southern California surface waters in warm years (Wrobel and Mills 1998).

Velamen parallelum. The parentheses around author and date were mistakenly omitted in the first edition; the species was originally described as *Vexillum.*

Beroe abyssicola. This cold temperate species is very likely present both in the Arctic and in the North Atlantic, but we can find no citations of its presence there.

Beroe forskalii. This species is known from coastal California and Washington (Wrobel and Mills 1998; Haddock and Case 1999).

Beroe gracilis. This species is known from coastal California and Washington (Wrobel and Mills 1998; Haddock and Case 1999).

Beroe mitrata. The date of publication is 1907, not 1908, as in the first edition.

Beroe ovata. The author's name was misspelled in the first edition, with an acute accent instead of the correct grave accent.

Appendix 2: Endangered and Threatened Cnidarians of North America

On 18 January 1990, several orders of Cnidaria were added to Appendix II of CITES (the Convention on International Trade in Endangered Species). Regarding North American shallow-water (0–200 m) cnidarians, this affected 127 species of stony corals (Scleractinia), 19 species of black coral (Antipatharia), 20 species of the athecate hydroid family Stylasteridae (stylasterids), and 2 species of the athecate hydroid family Milleporidae (fire corals). Some of these species had been on CITES' Appendix II prior to 1990, but the 1990 listing was all-inclusive for these taxa. Contrary to the connotation of the title of the treaty, placement of a species on Appendix II does not mean that it is endangered or threatened; rather, the listing means that it may become so unless its international trade is regulated. Thus, any specimen pertaining to any of these orders and families that crosses an international border must be accompanied by an export permit from the country of origin and an import permit from the country of destination. These permits may be obtained from the management authorities of the respective countries. In the United States, permits are issued by the U.S. Fish and Wildlife Service, Office of Management Authority, 4401 North Fairfax Drive, Room 420 C, Arlington, Virginia 22203. The addresses of the management authorities for all 134 signatory countries are listed on the Internet (http://international.fws.gov/cites/mgmtaut1.html). Import of CITES-protected specimens must also be done through designated ports and with advanced notification. Certain noncommercial institutions, such as museums and universities, may apply for a scientific exchange certificate, which allows a more expeditious exchange of specimens across international borders, but only among those institutions holding the exemption. There are more than a thousand exempt institutions worldwide. Fossil Cnidaria are not regulated by CITES. Cnidarians sent to or from U.S. territories, such as Guam, Puerto Rico, the Commonwealth of the Northern Marianas (14 islands, including Saipan, Rota, and Tinian), the U.S. Virgin Islands, and America Samoa do not require export-import permits. An informative review of the nature and quantity of international trade in CITES-regulated cnidarians through 1997 is given by Green and Shirley (1999). More information about CITES regulations can be found at the following website: http://international.fws.gov/cites/cites.html and http://www.cites.org.

The Endangered Species Act of 1973 authorized an official list of endangered and threatened wildlife. Responsibility for adding a marine species to this list is under the jurisdiction of the National Oceanographic and Atmospheric Administration (NOAA), in the Department of Commerce. No North American cnidarians are on this list; however, the *Federal Register* of 23 June 1999 (64(120):33466-33468) listed two stony corals species (*Acropora palmata* and *A. cervicornis*) as candidates for the list. Candidate species are not considered endangered or threatened, but as worthy of study to ascertain whether they should eventually be placed on the list.

In addition to these federal regulations, the state of Florida prohibits the collection of all stony coral (Scleractinia), fire coral (*Millepora* spp.), and two gorgonacean sea fans (*Gorgonia flabellum* and *G. ventalina*) from its waters.

Green, E. P., and F. Shirley. 1999. The global trade in coral. World Conservation Monitoring Centre. World Conservation Press, Cambridge, UK.

Appendix 3
References

This appendix lists all literature cited (marked with an asterisk or cross) used to document the changes and additions to the species list (Part I).

Cubozoa and Scyphozoa

Brodeur, R. D., C. E. Mills, J. E. Overland, G. E. Walters, and J. D. Schumacher. 1999. Evidence for a substantial increase in gelatinous zooplankton in the Bering Sea, with possible links to climate change. Fisheries Oceanography 8:296–306.

Clark, H. J. 1863. Prodromus of the history, structure, and physiology of the order Lucernariae. Journal of the Boston Society of Natural History 7:531–567.

Eckelbarger, K. J., and R. J. Larson. 1993. Ultrastructural study of the ovary of the sessile scyphozoan, *Haliclystus octoradiatus* (Cnidaria: Stauromedusae). Journal of Morphology 218:225–236.

Greenberg, N., R. L. Garthwaite, and D. C. Potts. 1996. Allozyme and morphological evidence for a newly introduced species of *Aurelia* in San Francisco Bay, California. Marine Biology 125: 401–410.

Gwilliam, G. F. 1956. Studies on West Coast Stauromedusae. Ph.D. Dissertation, University of California, Berkeley.

Hirano, Y. M. 1986. Species of Stauromedusae from Hokkaido, with notes on their metamorphosis. Journal of the Faculty of Science, Hokkaido University, Series 6 (Zoology) 24:182–201.

Hirano, Y. M. 1997. A review of a supposedly circumboreal species of Stauromedusa, *Haliclystus auricula* (Rathke, 1806). Proceedings of the 6th International Conference on Coelenterate Biology 1995:247–252.

Hyman, L. H. 1940. Observations and experiments on the physiology of medusae. Biological Bulletin 79:282–296.

Kramp, P. L. 1961. Synopsis of the medusae of the world. Journal of the Marine Biological Association of the United Kingdom 40:1–469.

Larson, R. J. 1976. Marine flora and fauna of the northeastern United States. Cnidaria: Scyphozoa. NOAA (National Oceanic and Atmospheric Administration) Technical Report NMFS (National Marine Fisheries Service) Circular 397:1–17.

Larson, R. J. 1986. Pelagic Scyphomedusae (Scyphozoa: Coronatae and Semaeostomeae) of the Southern Ocean. Antarctic Research Series 41(3):59–165.

Larson, R. J. 1988. *Kyopoda lamberti* gen. nov., sp. nov., an atypical Stauromedusa (Scyphozoa, Cnidaria) from the eastern Pacific, representing a new family. Canadian Journal of Zoology 66:2301–2303.

Larson, R. J., and A. C. Arneson. 1990. Two medusae new to the coast of California: *Carybdea marsupialis* (Linnaeus, 1758), a Cubomedusa and *Phyllorhiza punctata* von Lendenfeld, 1884, a rhizostome scyphomedusa. Bulletin of the Southern California Academy of Sciences 89:130–136.

Larson, R. J., and D. G. Fautin. 1989. Stauromedusae of the genus *Manania* (=*Thaumatoscyphus*) (Cnidaria, Scyphozoa) in the northeast Pacific, including descriptions of new species *Manania gwilliami* and *Manania handi*. Canadian Journal of Zoology 67:1543–1549.

Martin, J. W., L. A. Gershwin, J. W. Burnett, D. G. Cargo, and D. A. Bloom. 1997. *Chrysaora achlyos*, a remarkable new species of scyphozoan from the eastern Pacific. Biological Bulletin 193:8–13.

Matsumoto, G. I. 1995. Observations of the anatomy and behaviour of the cubozoan *Carybdea rastonii* Haacke. Marine and Freshwater Behaviour and Physiology 26:139–148.

Mayer, A. G. 1910. Medusae of the world. III. Scyphomedusae. Carnegie Institution, Washington, D.C.

Mianzan, H. W., and P. F. S. Cornelius. 1999. *Cubomedusae and Scyphomedusae.* Pages 513–559 *in* D. Boltovskoy, editor. South Atlantic zooplankton. Backhuys Publishers, Leiden, The Netherlands.

Mills, C. E. 1996. *Additions and corrections to the keys to Hydromedusae, Hydroid polyps, Siphonophora, Stauromedusan Scyphozoa, Actiniaria, and Ctenophora.* Pages 487–491 *in* E. N. Kozloff, editor. Marine invertebrates of the Pacific Northwest, with revisions and corrections. University of Washington Press, Seattle.

Mills, C. E. 1999–2000. Stauromedusae: list of all valid species names. Electronic internet document available at http://faculty.washington.edu/cemills/Staurolist.html. Published by the author, web page established October 1999, last viewed 6 July 2000.

Mills, C. E., R .J. Larson, and M. J. Youngbluth. 1987. A new species of coronate scyphomedusa from the Bahamas, *Atorella octogonos.* Bulletin of Marine Science 40:423–427.

Ralph, P. M. 1960. *Tetraplatia*, a coronate scyphomedusan. Proceedings of the Royal Society, Series B, 152:263–281.

Russell, F. S. 1970. The medusae of the British Isles. II. Pelagic Scyphozoa. Cambridge University Press, Cambridge, UK.

Wrobel, D., and C. E. Mills. 1998. Pacific coast pelagic invertebrates: a guide to the common gelatinous animals. Sea Challengers and the Monterey Bay Aquarium, Monterey, California.

Anthoathecatae and Leptothecatae (hydroids)

Agassiz, L. 1862. Contributions to the natural history of the United States of America. Vol. IV. Little, Brown, Boston.

Allman, G. J. 1876. Diagnoses of new genera and species of Hydroida. Journal of the Linnean Society, Zoology 12:251–284.

Boero F., and J. Bouillon. 1993. *Fraseroscyphus sinuosus* n. gen. (Cnidaria, Hydrozoa, Leptomedusae, Sertulariidae), an epiphytic hydroid with a specialised clinging organ. Canadian Journal of Zoology 71:1061–1064.

Boero, F., J. Bouillon, and C. Gravili. 1991. The life cycle of *Hydrichthys mirus* (Cnidaria: Hydrozoa: Anthomedusae: Pandeidae). Zoological Journal of the Linnean Society 101:189–199.

Boero, F., J. Bouillon, and C. Gravili. 2000. A survey of Zanclea, Halocoryne and Zanclella (Cnidaria, Hydrozoa, Anthomedusae, Zancleidae) with description of new species. Italian Journal of Zoology 67:93–124.

Boero, F., J. Bouillon, and S. Kubota. 1997. The medusae of some species of *Hebella* Allman, 1888, and *Anthohebella* gen. nov. (Cnidaria, Hydrozoa, Lafoeidae), with a world synopsis of species. Zoologische Verhandelingen, Leiden 310:1–53.

Boero, F., J. Bouillon, and S. Piraino. 1998. Heterochrony, generic distinction and phylogeny in the family Hydractiniidae (Hydrozoa: Cnidaria). Zoologische Verhandelingen, Leiden 323:25–36.

Boero, F., and C. Hewitt. 1992. A hydrozoan, *Zanclella bryozoophila* n.g., n.sp. (Zancleidae), symbiotic with a bryozoan, with a discussion of the Zancleoidea. Canadian Journal of Zoology 70:1645–1651.

Bogle, M. A. 1975. A review and preliminary revision of the Aglaopheniinae (Hydroida: Plumulariidae) of the tropical western Atlantic. Master's Thesis, University of Miami, Coral Gables.

Bouillon, J. 1995. Cnidaires. Cténaires. Traité de zoologie, Vol. 3, Fascicule 2. Paris, Masson.

Brinckmann-Voss, A. 1970. Anthomedusae/Athecatae (Hydrozoa, Cnidaria) of the Mediterranean. Part I. Capitata. Fauna e Flora del Golfo di Napoli, Monograph 39:1–96.

Brinckmann-Voss, A. 1989. *Sarsia cliffordi* n.sp. (Cnidaria, Hydrozoa, Anthomedusae) from British Columbia, with distribution records and evaluation of related species. Canadian Journal of Zoology 67:685–591.

Brinckmann-Voss, A. 1996. Seasonality of hydroids (Hydrozoa, Cnidaria) from an intertidal pool and adjacent subtidal habitats at Race Rocks, off Vancouver Island, Canada. Scientia Marina 60:89–97.

Brinckmann-Voss, A., D. M. Lickey, and C. E. Mills. 1993. *Rhysia fletcheri* (Cnidaria, Hydrozoa, Rhysiidae), a new species of colonial hydroid from Vancouver Island (British Columbia, Canada) and the San Juan Archipelago (Washington, U.S.A.). Canadian Journal of Zoology 71:401–406.

Broch, H. 1916. Hydroida (Part I). Danish Ingolf-Expedition 5(6):1–66.

Broch, H. 1918. Hydroida (Part II). Danish Ingolf-Expedition 5(7):1–205.

Buss, L. W., and P. O. Yund. 1989. A sibling species group of *Hydractinia* in the northern United States. Journal of the Marine Biological Association of the United Kingdom 69:857–874.

Cairns, S. D. 1984. A generic revision of the Stylasteridae (Coelenterata: Hydrozoa). Part 2: phylogenetic analysis. Bulletin of Marine Science 35:38–53.

Cairns S. D., and H. Zibrowius. 1992. Revision of the northeast Atlantic and Mediterranean Stylasteridae (Cnidaria: Hydrozoa). Mémoires du Muséum National d'Histoire Naturelle (A, Zoologie) 153:1–136.

Calder, D. R. 1970. Thecate hydroids from the shelf waters of northern Canada. Journal of the Fisheries Research Board of Canada 27:1501–1547.

Calder, D. R. 1975. Biotic census of Cape Cod Bay: hydroids. Biological Bulletin 149:287–315.

Calder, D. R. 1988. Shallow-water hydroids of Bermuda. The Athecatae. Royal Ontario Museum Life Sciences Contributions 148:1–107.

Calder, D. R. 1991. Shallow-water hydroids of Bermuda. The Thecatae, exclusive of Plumularioidea. Royal Ontario Museum Life Sciences Contributions 154:1–140.

Calder, D. R. 1997. Shallow-water hydroids of Bermuda. Superfamily Plumularioidea. Royal Ontario Museum Life Sciences Contributions 161:1–85.

Calder, D. R., and L. D. Stephens. 1997. The hydroid research of American naturalist Samuel F. Clarke, 1851–1928. Archives of Natural History 24:19–36.

Calder, D. R., L. D. Stephens, and A. E. Sanders. 1992. Comments on the date of publication of John McCrady's hydrozoan paper *Gymnopthalmata of Charleston Harbor*. Bulletin of Zoological Nomenclature 49:287–288.

Calder, D. R., and W. Vervoort. 1998. Some hydroids (Cnidaria: Hydrozoa) from the Mid-Atlantic Ridge, in the North Atlantic Ocean. Zoologische Verhandelingen, Leiden 319:1–65.

Calkins, G. N. 1899. Some hydroids from Puget Sound. Proceedings of the Boston Society of Natural History 28:333–368.

Clark, S. F. 1876. The hydroids of the Pacific coast of the United States, south of Vancouver Island. With a report upon those in the museum of Yale College. Transactions of the Connecticut Academy of Arts and Sciences 3:249–264.

Cornelius, P. F. S. 1975. A revision of the species of Lafoeidae and Haleciidae (Coelenterata: Hydroida) recorded from Britain and nearby seas. Bulletin of the British Museum (Natural History), Zoology 28:375–426.

Cornelius, P. F. S. 1982. Hydroids and medusae of the family Campanulariidae recorded from the eastern North Atlantic, with a world synopsis of genera. Bulletin of the British Museum (Natural History), Zoology 42:37–148.

Cornelius, P. F. S. 1992. Medusa loss in leptolid Hydrozoa (Cnidaria), hydroid rafting, and abbreviated life-cycles among their remote-island faunae: an interim review. Scientia Marina 56:245–261.

Cornelius, P. F. S. 1995a. North-west European thecate hydroids and their medusae. Part 1. Introduction, Laodiceidae to Haleciidae. Synopses of the British Fauna (New Series) 50:1–347.

Cornelius, P. F. S. 1995b. North-west European thecate hydroids and their medusae. Part 2. Sertulariidae to Campanulariidae. Synopses of the British Fauna (New Series) 50:1–386.

Edwards, C. 1972. The hydroids and the medusae *Podocoryne areolata*, *P. borealis* and *P. carnea*. Journal of the Marine Biological Association of the United Kingdom 52:97–144.

Edwards, C. 1973. The medusa *Modeeria rotunda* and its hydroid *Stegopoma fastiatum*, with a review of *Stegopoma* and *Stegolaria*. Journal of the Marine Biological Association of the United Kingdom 53:573–600.

Edwards, C., and S. M. Harvey. 1975. The hydroids *Clava multicornis* and *Clava squamata*. Journal of the Marine Biological Association of the United Kingdom 55:879–886.

Fraser, C. M. 1911. The hydroids of the west coast of North America with special reference to those of the Vancouver Island region. Bulletin from the Laboratories of Natural History of the State University of Iowa 6:1–91.

Fraser, C. M. 1937. Hydroids of the Pacific coast of Canada and the United States. University of Toronto Press, Toronto.

Fraser, C. M. 1940. Some hydroids from the California coast, collected in 1939. Transactions of the Royal Society of Canada, 3rd Series, Section 5, 34:39–44.

Fraser, C. M. 1944. Hydroids of the Atlantic coast of North America. University of Toronto Press, Toronto.

Fraser, C. M. 1948. Hydroids of the Allan Hancock Pacific Expeditions since March, 1938. Allan Hancock Pacific Expeditions 4:179–335.

Gibbons, M. J., and J. S. Ryland. 1989. Intertidal and shallow water hydroids from Fiji. I. Athecata to Sertulariidae. Memoirs of the Queensland Museum 27:377–432.

Hirohito, The Sh?wa Emperor. 1988. The hydroids of Sagami Bay. Biological Laboratory, Imperial Household, Tokyo.

Hirohito, The Sh?wa Emperor. 1995. The hydroids of Sagami Bay II. Thecata. Biological Laboratory, Imperial Household, Tokyo.

Kramp, P. L. 1961. Synopsis of the medusae of the world. Journal of the Marine Biological Association of the United Kingdom 40:1–469.

Lees, D. C. 1968. The addition of the hydroid *Cladocoryne floccosa* to the California marine fauna. *Bulletin of the Southern California Academy of Science* 67:59–64.

Lees, D. C. 1986. Marine hydroid assemblages in soft-bottom habitats on the Hueneme Shelf off Southern California, and factors influencing hydroid distribution. Bulletin of the Southern California Academy of Science 85:102–119.

Lindner, A., and D. R. Calder. 2000. *Campanularia noliformis* McCrady, 1859 (currently *Clytia noliformis*, Cnidaria, Hydrozoa): proposed conservation of the specific name by the designation of a neotype. Bulletin of Zoological Nomenclature 57:140–143.

Medel, M. D., and W. Vervoort. 1998. Atlantic Thyroscyphidae and Sertulariidae (Hydrozoa, Cnidaria) collected during the CANCAP and Mauritania-II expeditions of the National Museum of Natural History, Leiden, The Netherlands. Zoologische Verhandelingen 320:1–85.

Millard, N. A. H. 1975. Monograph on the Hydroida of southern Africa. Annals of the South African Museum 68:1–513.

Mills, C. E., and R. L. Miller. 1987. *Hydroid polyps*. Pages 44–62 *in* E. N. Kozloff, editor. Marine invertebrates of the Pacific Northwest. University of Washington Press, Seattle.

Namikawa, H. 1991. A new species of the genus *Stylactaria* (Cnidaria, Hydrozoa) from Hokkaido, Japan. Zoological Science 8:805–812.

Naumov, D. V. 1960. Gidroidy i gidromeduzy morskikh, solonovatovodnykh i presnovodnykh basseinov SSSR. Akademiya Nauk SSSR, Opred. Faune SSSR 70:1–626.

Nutting, C. C. 1896. Notes on Plymouth hydroids. Journal of the Marine Biological Association of the United Kingdom (New Series) 4:146–154.

Nutting, C. C. 1915. American hydroids. Section III. The Campanularidae and Bonneviellidae. Smithsonian Institution, United States National Museum Special Bulletin 4(3):1–126.

Petersen, K. W. 1990. Evolution and taxonomy in capitate hydroids and medusae (Cnidaria: Hydrozoa). Zoological Journal of the Linnean Society 100:101–231.

Rees, W. J. 1938. Observations on British and Norwegian hydroids and their medusae. Journal of the Marine Biological Association of the United Kingdom 23:1–42.

Rees, W. J. 1939. A revision of the genus *Campanulina* van Beneden, 1847. Annals and Magazine of Natural History (11)3:433–447.

Rees, W. J. 1956. A revision of the hydroid genus *Perigonimus* M. Sars, 1846. Bulletin of the British Museum (Natural History), Zoology 3:337–350.

Rees, W. J. 1957. Evolutionary trends in the classification of capitate hydroids and medusae. Bulletin of the British Museum (Natural History), Zoology 4:453–534.

Rees, W. J., and W. Vervoort. 1987. Hydroids from the John Murray Expedition to the Indian Ocean, with revisory notes on *Hydrodendron*, *Abietinella*, *Cryptolaria* and *Zygophylax* (Cnidaria: Hydrozoa). Zoologische Verhandelingen, Leiden 237:1–209.

Russell, F. S. 1953. The medusae of the British Isles. Anthomedusae, Leptomedusae, Limnomedusae, Trachymedusae and Narcomedusae. Cambridge University Press, Cambridge.

Schuchert, P. 1996. The marine fauna of New Zealand: athecate hydroids and their medusae (Cnidaria: Hydrozoa). New Zealand Oceanographic Institute Memoir 106:1–159.

Schuchert, P. 1997. Review of the family Halopterididae (Hydrozoa, Cnidaria). Zoologische Verhandelingen 309:1–162.

Sheiko, O., and S. Stepanjants. 1997. Meduzoa (Cnidaria: Anthozoa excepted) from the Commander Islands, faunistic composition and biogeography. Proceedings of the 6th International Conference on Coelenterate Biology 1995:437–445.

Stechow, E. 1920. Neue Ergebnisse auf dem Gebiete der Hydroidenforschung. Sitzungsberichte der Gesellschaft für Morphologie und Physiologie in München 31:9–45.

Vannucci, M., and W. J. Rees. 1961. A revision of the genus *Bougainvillia* (Anthomedusae). Boletim do Instituto Oceanográfico 11(2):57–100.

Vervoort, W. 1946. Hydrozoa (C 1). A. Hydropolypen. Fauna van Nederland, Aflevering 14:1–336.

Vervoort, W. 1964. Note on the distribution of *Garveia franciscana* (Torrey, 1902) and *Cordylophora caspia* (Pallas, 1771) in the Netherlands. Zoologische Mededelingen, Leiden 39:125–146.

Vervoort, W. 1972. Hydroids from the Theta, Vema and Yelcho cruises of the Lamont-Doherty Geological Observatory. Zoologische Verhandelingen, Leiden 92:1–247.

Anthoathecatae and Leptothecatae (hydromedusae)

Alavariño, A. 1999. Hidromedusas: abundancia batimétrica diurna-nocturna y estacional en aguas de California y Baja California, y estudio de las especies en el Pacífico oriental y otras regiones. Revista de Biologia Marina y Oceanografia 34(1):1–90.

Arai, M. N., and A. Brinckmann-Voss. 1980. Hydromedusae of British Columbia and Puget Sound. Canadian Bulletin of Fisheries and Aquatic Sciences 204:1–192.

Arai, M. N., and A. Brinckmann-Voss. 1983. A new species of *Amphinema: Amphinema platyhedos* n. sp. (Cnidaria, Hydrozoa, Pandeidae) from the Canadian West Coast. Canadian Journal of Zoology 61:2179–2182.

Bigelow, H. B. 1909. Reports on the scientific results of the expedition to the Eastern tropical Pacific, in charge of Alexander Agassiz, by the U.S. Fish Commission Steamer "Albatross." XVI. The medusae. Memoirs of the Museum of Comparative Zoology at Harvard College 37:1–243.

Boero, F., and J. Bouillon. 1989. The life cycles of *Octotiara russelli* and *Stomatoca atra* (Cnidaria, Anthomedusae, Pandeidae). Zoologica Scripta 18(1):1–7.

Boero, F., C. Gravili, F. Denitto, and M. Miglietta. 1997. The rediscovery of *Codonorchis octaedrus* (Hydroidomedusae, Anthomedusae, Pandeidae), with an update of the Mediterranean hydroidomedusan biodiversity. Italian Journal of Zoology 64:359–365.

Bouillon, J. 1978. Hydroméduses de la mer de Bismarck (Papouasie, Nouvelle-Guinée). I.- Anthomedusae Capitata (Hydrozoa-Cnidaria). Cahiers de Biologie Marine 19:249–297.

Bouillon, J. 1985. Essai de classification des Hydropolypes-Hydroméduses (Hydrozoa-Cnidaria). Indomalayan Zoology 2(1):29–243.

Bouillon, J. 1999. *Hydromedusae.* Pages 385–465 *in* D. Boltovsky, editor. South Atlantic zooplankton. Backhuys Publishers, Leiden, The Netherlands.

Brinckmann-Voss, A. 1970. Anthomedusae/Athecatae (Hydrozoa, Cnidaria) of the Mediterranean. Part I. Capitata. Fauna e Flora del Golfo di Napoli, Monograph 39:1–96.

Brinckmann-Voss, A. 2000. The hydroid and medusa of *Sarsia bella* sp. nov. (Hydrozoa, Anthoathecatae, Corynidae), with a correction of the "life cycle" of *Polyorchis penicillatus* (Eschscholtz). Scientia Marina, 64(supplement 1):189–195.

Brinckmann-Voss, A., and M. N. Arai. 1998. Further notes on Leptolida (Hydrozoa: Cnidaria) from Canadian Pacific Waters. Commemorative volume for the 80th birthday of Willem Vervoort in 1997. Zoologische Verhandelingen, Leiden 323:37–68.

Calder, D. R. 1970. Thecate hydroids from the shelf waters of northern Canada. Journal of the Fisheries Research Board of Canada 27:1501–1547.

Calder, D. R. 1971. Hydroids and hydromedusae from Southern Chesapeake Bay. Virginia Institute of Marine Science Special Papers in Marine Science 1:1–125.

Calder, D. R. 1991. Shallow-water hydroids of Bermuda. The Athecatae. Royal Ontario Museum Life Sciences Contributions 154:1–140.

Cornelius, P. F. S. 1995. North-west European thecate hydroids and their medusae. Part I, II. Synopses of the British Fauna (New Series) 50:1–347 (part I); 1–386 (part II).

Foerster, R. E. 1923. The hydromedusae of the west coast of North America, with special reference to those of the Vancouver Island region. Contributions to Canadian Biology (New Series) 1:219–277.

Kozloff, E. N. 1996. Marine Invertebrates of the Pacific Northwest. University of Washington Press, Seattle and London.

Kramp, P. 1961. Synopsis of the medusae of the world. Journal of the Marine Biological Association of the United Kingdom 40:1–469.

Mayer, A. G. 1910. Medusae of the World. Carnegie Institute of Washington Publications (reprinted in 1977 by Asher & Co., B.V., Amsterdam).

Mills, C. E. 2000. The life cycle of *Halimedusa typus* with discussion of other species closely related to the family Halimedusidae (Hydrozoa, Capitata, Anthomedusae). Scientia Marina, 64(supplement 1):97–106.

Mills, C. E,. and J. Rees, 2000. New observations and corrections concerning the trio of invasive hydromedusae *Maeotias marginata* (= *M. inexpectata*), *Blackfordia virginica*, and *Moerisia sp.* in the San Francisco estuary. Scientia Marina, 64(supplement 1):151–155.

Petersen, K. W. 1990. Evolution and taxonomy in capitate hydroids and medusae (Cnidaria: Hydrozoa). Zoological Journal of the Linnean Society 100:101–231.

Rees, J. 2000. A pandeid hydrozoan, *Amphinema sp.*, new and probably introduced to central California: life history, morphology, distribution, and systematics. Scientia Marina, 64(supplement 1):165–172.

Rees, J. T., and L. A. Gershwin. 2000. Non-indigenous hydromedusae in California's upper San Francisco Estuary: life cycles, distribution, and potential environmental impacts. Scientia Marina, 64(supplement 1):73–86.

Schuchert, P. 1996. The marine Fauna of New Zealand: athecate hydroids and their medusae (Cnidaria: Hydrozoa). New Zealand Oceanographic Institute Memoirs 106:1–159.

Stepanjants, S. D., A. L. Lobanov, and M. B. Dianov. 1999. New approaches to the search for interspecies diagnostic differences within the genus *Obelia.* Analysis of the growth and development of free medusae of *Obelia* spp. from different areas of the world Ocean (under laboratory conditions). Zoosystematica Rossica Suppl. No. 1:34–50.

Sugisaki, H., R. Brodeur, and J. M. Napp. 1998. Summer distribution and Macrozooplankton in the Western Gulf of Alaska and Southeastern Bering Sea. Memoirs of the Faculty of Fisheries Hokkaido University 45(1):96–112.

Thiel, M. E. 1932. Die Hydromedusenfauna des Nördlichen Eismeeres in tiergeographischer Betrachtung. Archiv Naturgeschichte N.F. 1:435–514.

Uchida, Th. 1938. Medusae in Onagawa Bay and its vicinity. Science Report Tohoku University 13(4):47–58.

Werner, B. 1984. Stamm Cnidaria. Pages 11–305 *in* H.-E. Gruner, editor. Lehrbuch der Speciellen Zoologie 2. Stuttgart, Gustav Fisher Verlag.

Wrobel, D., and C. E. Mills. 1998. Pacific coast pelagic invertebrates: a guide to the common gelatinous animals. Sea Challengers and Monterey Aquarium Publication.

Limnomedusae

Acker, T.S. 1976. Craspedacusta sowerbyi: an analysis of an introduced species. Pages 219–226 *in* G. O. Mackie, editor. Coelenterate ecology and behavior. Plenum Press, New York.

Bushnell, J. H., and T. W. Porter. 1967. Transactions of the American Microscopal Society 86(1):22–27.

Narcomedusae

Alvariño, A. 1999. Hidromedusas: abundancia batimétrica diurna-nocturna y estacional en aguas de California y Baja California, y estudio de las especies en el Pacifico oriental y otras regiones. Revista de Biologia Marina y Oceanografia 341–390.

Alvariño, A., and C. A. Kimbrell. 1987. Abundance of zooplankton species in California coastal waters during April 1981, February 1982, March 1984, March 1985. NOAA Technical Memoir, NMFS, NOAA-TM-NMFS-SWFC-74:1–59.

Arai, M. N., M. J. Cavey, and B. A. Moore. 2000. Morphology and distribution of a deep-water Narcomedusa (Solmarisidae) from the north-east Pacific. Scientia Marina, supplement 1:55–62.

Arai, M. N., G. A. McFarlane, M. W. Saunders, and G. M. Mapstone. 1993. Spring abundance of medusae, ctenophores, and siphonophores off southwest Vancouver Island: possible competition or predation on sablefish larvae. Canadian Technical Report of Fisheries and Aquatic Sciences 1939:1–37.

Bouillon, J. 1995. *Classe des Hydrozoaires (Hydrozoa Owen, 1843).* Pages 29–416 *in* Traité de Zoologie, Anatomie, Systématique, Biologie III Cnidaires, Cténaires 2. Masson, Paris.

Haddock, S. H. D., and J. F. Case. 1999. Bioluminescence spectra of shallow and deep-sea gelatinous zooplankton: ctenophores, medusae and siphonophores. Marine Biology 133:571–582.

Hanelt, B., D. van Schyndel, C. M. Adema, L. A. Lewis, and E. S. Loker. 1996. The phylogenetic position of *Rhopalura ophiocomae* (Orthonectida) based on 18S ribosomal DNA sequence analysis. Marine Biology Evolution 13:1187–1191.

Kent, M. L., L. Margolis, and J. O. Corliss. 1994. The demise of a class of protists: taxonomic and nomenclatural revisions proposed for the protist phylum Myxozoa Grassé, 1970. Canadian Journal of Zoology 72:932–937.

Kim, J., W. Kim, and C. W. Cunningham. 1999. A new perspective on lower metazoan relationships from 18S rDNA sequences. Marine Biology Evolution 16:423–427.

Lom, J., and I. Dykova. 1997. Ultrastructural features of the actinosporean phase of myxosporea (Phylum Myxozoa): a comparative study. Acta Protozoology 36:83–103.

Raikova, E. V. 1973. Life cycle and systematic position of *Polypodium hydriforme* Ussov (Coelenterata), a cnidarian parasite of the eggs of Acipenseridae. Publications of the Seto Marine Biology Laboratory 20:165–173.

Raikova, E. V. 1988. On the systematic position of Polypodium hydriforme Ussov, 1885. Pages 116–122 *in* V. M. Kotun and S. D. Stepanjants, editors. Porifera and Cnidaria: modern and perspective investigations (in Russian, English abstract). USSR Academy of Sciences, Moscow.

Raikova, E. V. 1994. Life cycle, cytology, and morphology of *Polypodium hydriforme*, a coelenterate parasite of the eggs of acipensieriform fishes. Journal of Parasitology 80:1–22.

Robison, B., and J. Connor. 1999. The deep sea. Bay Aquarium Press, Monterey, California.

Siddall, M. E., D. S. Martin, D. Bridge, S. S. Desser, and D. K. Cone. 1995. The demise of a phylum of protists: phylogeny of Myxozoa and other parasitic Cnidaria. Journal of Parasitology 81:961–967.

Siddall, M. E., and M. F. Whiting. 1999. Long-branch abstractions. Cladistics 15:9–24.

Widder, E. A., S. A. Bernstein, D. F. Bracher, J. F. Case, K. R. Reisenbichler, J. J. Torres, and B. H. Robison. 1989. Bioluminescence in the Monterey Submarine Canyon: image analysis of video recordings from a midwater submersible. Marine Biology 100:541–551.

Trachymedusae

Bouillon, J., and T. J. Barnett. 1999. The marine fauna of New Zealand: Hydromedusae (Cnidaria: Hydrozoa). NIWA Biodiversity Memoir 113:1–136.

Gili, J.-M., J. Bouillon, F. Pages, A. Palanques, P. Puig, and S. Heussner. 1998. Origin and biogeography of the deep-water Mediterranean Hydromedusae including the description of two new species collected in submarine canyons of northwestern Mediterranean. Scientia Marina 62:113–134.

Kramp, P. L. 1961. Synopsis of the medusae of the world. Journal of the Marine Biological Association of the United Kingdom 40:1–469.

Kramp, P. L. 1965. The Hydromedusae of the Pacific and Indian Oceans. Dana-Report 63:1–162.

Larson, R. J., and G. H. Harbison. 1990. Medusae from McMurdo Sound, Ross Sea including the descriptions of two new species, *Leuckartiara brownei* and *Benthocodon hyalinus*. Polar Biology 11:19–25

Larson, R. J., C. E. Mills, and G. R. Harbison. 1991. Western Atlantic midwater hydrozoan and scyphozoan medusae: *in situ* studies using manned submersibles. Hydrobiologia 216/217: 311–317.

Robison, B., and J. Connor. 1999. The deep sea. Monterey Bay Aquarium Press, Monterey, California.

Thuesen, E. V. 1993. *Vampyrocrossota childressi*, a new genus and species of black medusa from the bathypelagic zone off California (Cnidaria: Trachymedusae: Rhopalonematidae). Proceedings of the Biological Society of Washington 106(1): 190–194.

Thuesen, E. V., and J. J. Childress. 1994. Oxygen consumption rates and metabolic enzyme activities of oceanic California medusae in relation to body size and habitat depth. Biological Bulletin 187:84–98.

Wrobel, D., and C. E. Mills. 1998. Pacific coast pelagic invertebrates: a guide to the common gelatinous animals. Sea Challengers and the Monterey Bay Aquarium, Monterey, California.

Siphonophorae

Alvariño, A. 1967. Bathymetric distribution of Chaetognatha, Siphonophorae, Medusae and Ctenophorae off San Diego, California. Pacific Science 21:474–485.

Alvariño, A. 1980. The relation between the distribution of zooplankton predators and anchovy larvae. CalCOFI Reports 21:150–160.

Alvariño, A. 1985. Distribucion batimetrica de especies del genera *Lensia* en aguas de California y Baja California (Diphyidae, Siphonophorae, Coelenterata). Investigionas Marinas CICIMAR 2:59–80.

Alvariño, A. 1991. Abunancia y distribucion batimetrica diurna y nocturna de los sifonoforos durante las cuatro estaciones del año 1969, en aguas de California y Baja California. Inv. Mar. CICIMAR 6:1–37.

Arai, M. N., G. A. McFarlane, M. W. Saunders, and G. M. Mapstone. 1993. Spring abundance of medusae, ctenophores, and siphonophores off southwestern Vancouver Island: possible competition or predation on sablefish larvae. Canadian Technical Report of Fisheries and Aquatic Sciences 1939:1–37.

Bouillon, J., F. Boero, F. Cicogna, J.-M. Gili, and R. G. Hughes. 1992. Non-siphonophoran Hydrozoa: what are we talking about? Scientia Marina 56:279–284.

Gasca, R. 1999. Siphonophores (Cnidaria) and summer mesoscale features in the Gulf of Mexico. Bulletin of Marine Science 65:75–89.

Gasca, R., and E. Suárez. 1991. Nota sobre los sifonoforos (Cnidaria: Siphonophora) del Golfo de California (Agosto-Septiembre, 1977). Ciencia Pesquera 8:119–125.

Gasca, R., and E. Suárez. 1992. Sifonoforos (Cnidaria: Hydrozoa) de la zona sudoccidental de la peninsula de Baja California, en invierno y verano durante "El Niño" 1983. Revista de Investigación Científica 3:37–46.

Haddock, S. H. D., and J. F. Case. 1999. Bioluminescence spectra of shallow and deep-sea gelatinous zooplankton: ctenophores, medusae and siphonophores. Marine Biology 133:571–582.

Mackie, G. O., P. R. Pugh, and J. E. Purcell. 1987. Siphonophore biology. Advances in Marine Biology 24:97–262.

Mapstone, G. M. 1998. *Bargmannia lata*, an undescribed physonect siphonophore (Cnidaria, Hydrozoa) from Canadian Pacific waters. Zoologische Verhandelingen 323:141–147.

Margulis, R. Ya. 1978. The distribution of siphonophores in the western North Atlantic in summer of 1974. Vestnik Moskovskogo Universiteta 3:1–11 (in Russian).

Margulis, R. Ya. 1988. Revision of the subfamily Clausophyinae (Siphonophora, Diphyidae). Zoologicheskii Zhurnal 67:1269–1281 (in Russian).

Margulis, R. Ya. 1994. Revision of the genus *Rosacea* (Cnidaria, Siphonophora, Calycophorae, Prayidae, Prayinae). Hydrobiological Journal 31:33–50. Translation of Zoologicheskii Zhurnal 73:15–28 (1995).

Margulis, R. Y., and A. L. Vereshchaka. 1994. Siphonophores from the North Pacific. Trudy Instituta Okeanologii 131:76–89 (in Russian).

Moser, F. 1925. Die Siphonophoren der Deutschen Südpolar-Expedition, 1901–03. Deutsche Südpolar-Expedition 17(Zool. 9):1–541.

Pugh, P. R. 1992a. A revision of the sub-family Nectopyramidinae (Siphonophora, Prayidae). Philosophical Transactions of the Royal Society of London (B)335:281–322.

Pugh, P. R. 1992b. The status of the genus *Prayoides* (Siphonophora: Prayidae). Journal of the Marine Biological Association of the United Kingdom 72:895–909.

Pugh, P. R. 1992c. *Desmophyes haematogaster*, a new species of prayine siphonophore (Calycophorae, Prayidae). Bulletin of Marine Science 50:89–96.

Pugh, P. R. 1998. A re-description of *Frillagalma vityazi* Daniel 1966 (Siphonophorae, Agalmatidae). Scientia Marina 62:233–245.

Pugh, P. R. 1999. A review of the genus *Bargmannia* Totton, 1954 (Siphonophorae, Physonecta, Pyrostephidae). Bulletin of the Natural History Museum, London (Zoology Series) 65:51–72.

Pugh, P. R., and G. R. Harbison. 1987. Three new species of prayine siphonophore (Calycophorae, Prayidae) collected by a submersible, with notes on related species. Bulletin of Marine Science 41:68–91.

Pugh, P. R., and F. Pagès. 1995. Is *Lensia reticulata* a diphyine species (Siphonophora, Calycophora, Diphyidae)? A re-description. Scientia Marina 59:181–192.

Pugh, P. R., and M. J. Youngbluth. 1988a. Two new species of prayine siphonophore (Calycophorae, Prayidae) collected by the submersibles *Johnson-Sea-Link I* and *II*. Journal of Plankton Research 10:637–657.

Pugh, P. R., and M. J. Youngbluth. 1988b. A new species of *Halistemma* (Siphonophora, Physonectae, Agalmidae) collected by submersible. Journal of the Marine Biological Association of the United Kingdom 68:1–14.

Totton, A. K. 1965a. A synopsis of the Siphonophora. British Museum (Natural History), London.

Totton, A. K. 1965b. A new species of *Lensia* (Siphonophora: Diphyidae) from the coastal waters of Vancouver, B.C.; and its comparison with *Lensia achilles* Totton and another new species *Lensia cordata*. Annals and Magazine of Natural History (13)8:71–76.

Antipatharia

Opresko, D. M. 1998. Three new species of Leiopathidae from southern Australia. Records of the South Australian Museum 31(1):99–111.

Ceriantharia

Agassiz, A. 1862. Proceedings of the Boston Society of Natural History 9:159.

Agassiz, A. 1863. On *Arachnactis brachiolata*, a species of floating *Actinia* found at Nahant, Massachusetts. Boston Journal of Natural History, 7:525–531.

Arai, M. N. 1971. *Pachycerianthus* (Ceriantharia) from British Columbia and Washington. Journal of the Fisheries Research Board of Canada 28:1677–1680.

Carlgren, O. 1924. Die Larven der Ceriantharien, Zoantharien und Actiniarien der Deutschen Tiefsee-Expedition mit einem Nachtrag zu den Zoantharien. Deutsche Tiefsee-Expedition 1898–1899 19(8):339–476.

Carlgren, O., and J. W. Hedgpeth. 1952. Actiniaria, Zoantharia and Ceriantharia from shallow water in the northwestern Gulf of Mexico. Publications of the Institute of Marine Science, University of Texas 2(2):141–172.

Leloup, E. 1962. Anthozoa: Ceriantharia: Larvae. Fisches Identif. Zooplancton 93:1–7.

Verrill, A. E. 1922. The Actiniaria of the Canadian Arctic Expeditions, with notes on interesting species from Hudson Bay and other Canadian localities. Report of the Canadian Arctic Expedition 1913–18 8G:89–165.

Widersten, B. 1998. On *Isodactylactis borealis*, a new species of cerianthid larvae. Helgolander Meeresunters 52:111–114.

Octocorallia

Bayer, F. M. 1957. Additional records of western Atlantic octocorals. Journal of the Washington Academy of Sciences 47(11):379–390.

Bayer, F. M. 1981. Key to the genera of Octocorallia exclusive of Pennatulacea, with diagnoses of new taxa. Proceedings of the Biological Society of Washington 94:902–947.

Bayer, F. M. 1982. Some new and old species of the primnoid genus *Callogorgia* Gray, with a revalidation of the related genus *Fanellia* Gray (Coelenterata: Anthozoa). Proceedings of the Biological Society of Washington 95(1):116–160.

Bayer, F. M. 1996. The gorgonacean genus *Arthrogorgia* (Octocorallia: Primnoidae). Proceedings of the Biological Society of Washington 109(4): 605–628.

Bayer, F. M., and M. Grasshoff. 1994. The genus group taxa of the family Ellisellidae, with the clarification of the genera established by J. E. Gray (Cnidaria: Octocorallia). Senckenbergiana Biologica 74(1–2):21–45.

Bayer, F. M., and M. Grasshoff. 1995. Two new species of the gorgonacean genus *Ctenocella* (Coelenterata: Anthozoa, Octocorallia) from deep reefs in the western Atlantic. Bulletin of Marine Science 56(2):625–652.

Deichmann, E. 1936. The Alcyonaria of the western part of the Atlantic Ocean. Memoirs of the Museum of Comparative Zoology at Harvard College 53:1–317.

Gabb, W. M. 1862. Description of two new species of Pennatulidae from the Pacific coast of the U.S. Proceedings of the California Academy of Natural Sciences 1(2):166–167.

Grasshoff, M. 1988. The genus *Leptogorgia* (Octocorallia: Gorgoniidae) in West Africa. Atlantide Report 14:91–147.

Grasshoff, M. 1992. Die Flachwasser-Gorgonarien von Europa und Westafrika (Cnidaria, Anthozoa). Courier Forschunginstitut Senckenberg 149:1–135.

Grasshoff, M. 1999. The shallow-water gorgonians of New Caledonia and adjacent islands (Coelenterata: Octocorallia). Senckenbergiana 78:1–121.

Gray, J. E. 1860. Revision of the Pennatulidae. Annals and Magazine of Natural History (3)5:20–25.

Humann, P. 1993. Reef coral identification: Florida, Caribbean, Bahamas, including marine plants. New World Publications, Jacksonville, Florida.

Kükenthal, W. 1924. Gorgonaria. Das Tierreich 47:1–478.

Marques-Paraense, A. C. 1996. Revisão taxonômica do gênero *Thesea* (Cnidaria: Octocorallia) do Atlântico Ocidental. Rio de Janeiro, Universidade Federal do Rio de Janeiro, Master's Thesis (in Portuguese).

Satterlie, R. A., and J. F. Case. 1979. Neurobiology of the gorgonian coelenterates, *Muricea californica* and *Lophogorgia chilensis*. I. Behavioural physiology. Journal of Experimental Biology 79:191–204.

Verrill, A. E. 1865. Synopsis of the polyps and corals of the North Pacific Exploring Expedition, under Commodore C. Ringgold and Captain John Rodgers, U.S.N., from 1853 to 1856. Collected by Dr. Wm. Stimpson, naturalist to the expedition.

With descriptions of some additional species from the west coast of North America. Proceedings of the Essex Institute, Salem 4:181–196.

Verseveldt, J., and L. P. van Ofwegen. 1992. New and redescribed species of *Alcyonium* Linnaeus, 1758 (Anthozoa: Alcyonacea). Zoologische Mededelingen (Leiden) 66(1–15):155–181.

Williams, G. C. 1995. Living genera of sea pens (Coelenterata: Octocorallia: Pennatulacea): Illustrated key and synopses. Zoological Journal of the Linnean Society 113:93–140.

Actiniaria, Zoanthidea, Corallimorpharia

Humann, P. 1992. Reef creature identification: Florida, Caribbean, Bahamas. New World Publications, Jacksonville, Florida.

Scleractinia

Breeze, H. D., S. Davis, and M. Butler. 1997. Distribution and status of deep sea corals off Nova Scotia. Marine Issues Committee Special Publication Number 1, Ecology Action Centre, Halifax.

Cairns, S. D. 1978. A checklist of the ahermatypic Scleractinia of the Gulf of Mexico, with the description of a new species. Gulf Research Reports 6(1):9–15.

Cairns, S. D. 1979. The deep-water Scleractinia of the Caribbean and adjacent waters. Studies on the Fauna of Curaçao and Other Caribbean Islands 57(180):1–341.

Cairns, S. D. 1982. Stony corals (Cnidaria: Hydrozoa, Scleractinia) of Carrie Bow Cay, Belize. Smithsonian Contributions to Marine Science 12:271–302.

Cairns, S. D. 1991a. A revision of the ahermatypic Scleractinia of the Galápagos and Cocos Islands. Smithsonian Contributions to Zoology 504:1–32.

Cairns, S. D. 1991b. Catalog of the type specimens of stony corals (Milleporidae, Stylasteridae, Scleractinia) in the National Museum of Natural History. Smithsonian Contributions to Zoology 514:1–59.

Cairns, S. D. 1994. Scleractinia of the temperate North Pacific. Smithsonian Contributions to Zoology 557:1–150.

Cairns, S. D. 1995. The marine fauna of New Zealand: Scleractinia (Cnidaria: Anthozoa). New Zealand Oceanographic Institute Memoir 103:1–210.

Cairns, S. D. 1997. A generic revision and phylogenetic analysis of the Turbinoliidae (Cnidaria: Scleractinia). Smithsonian Contributions to Zoology 591:1–55.

Cairns, S. D. 2000. A revision of the shallow-water Azooxanthellate Scleractinia of the western Atlantic. Studies of the Natural History of the Caribbean Region 75:1–231.

Cairns, S. D., D. M. Opresko, T. S. Hopkins, and W. W. Schroeder. 1993 (1994). New records of deep-water Cnidaria (Scleractinia and Antipatharia) from the Gulf of Mexico. Northeast Gulf Science 13(1):1–11.

Cairns, S. D., and H. Zibrowius 1997. Cnidaria Anthozoa: Azooxanthellate Scleractinia from the Philippine and Indonesian Regions. Mémoires du Muséum National d'Histoire Naturelle 172(2):27–243.

Hubbard, R., and J. W. Wells. 1986. Ahermatypic shallow-water Scleractinian corals of Trinidad. Studies on the Fauna of Curaçao and Other Caribbean Islands 68(211):121–147.

Humann, P. 1993. Reef coral identification: Florida, Caribbean, Bahamas, including marine plants. New World Publications, Jacksonville, Florida.

Messing, C. G. 1987. To the deep reef and beyond. Deep Ocean Society, Miami.

Rezak, R., T. J. Bright, and D. W. McGrail. 1985. Reefs and banks of the northwestern Gulf of Mexico: their geological, biological, and physical dynamics. John Wiley and Sons, New York.

Stolarski, J. 1996. Paleontological results of the Polish Antarctic Expeditions. Part II. Paleogene Corals from Seymour Island, Antarctic Peninsula. Palaeontologica Polonica 55:51–63.

Viada, S. T., and S. D. Cairns. 1987. Range extensions of ahermatypic Scleractinia in the Gulf of Mexico. Northeast Gulf Science 9(2):131–134.

Weil, E., and N. Knowlton. 1994. A multi-character analysis of the Caribbean coral *Montastraea annularis* (Ellis and Solander, 1786) and its two sibling species, *M. faveolata* (Ellis and Solander, 1786) and *M. franksi* (Gregory, 1895). Bulletin of Marine Science 55(1):151–175.

Wells, J. W. 1973. New and old scleractinian corals from Jamaica. Bulletin of Marine Science 23(1):16–55.

Zibrowius, H. 1980. Les Scléractiniaires de la Méditeranée et de l'Atlantique Nord-Oriental. Mémoires de l'Institut Océanographique, Monaco 11:1–284.

Zibrowius, H., and S. D. Cairns. 1992. Revision of the northeast Atlantic and Mediterranean Stylasteridae (Cnidaria: Hydrozoa). Mémoires du Muséum National d'Histoire Naturelle (A, Zoologie) 153:1–136.

Ctenophora

Bailey, T. G., M. J. Youngbluth, and G. P. Owen. 1995. Chemical composition and metabolic rates of gelatinous zooplankton from midwater and benthic boundary layer environments off Cape Hatteras, North Carolina, USA. Marine Ecology Progress Series 122:121–134.

Carré, C., and D. Carré. 1989. *Haeckelia bimaculata* sp. nov., une nouvelle espèce méditerranéenne de cténophore (Cydippida Haeckeliidae). Comptes Rendus de l'Académie des Sciences, Paris 308(Serie III):321–327.

Frank, T. M,. and E. A. Widder. 1997. The correlation of downwelling irradiance and staggered vertical migration patterns of zooplankton in Wilkinson Basin, Gulf of Maine. Journal of Plankton Research 19:1975–1991.

Haddock, S. H. D., and J. F. Case. 1995. Not all ctenophores are bioluminescent. Biological Bulletin 189:356–362.

Haddock, S. H. D., and J. F. Case. 1999. Bioluminescence spectra of shallow and deep-sea gelatinous zooplankton: ctenophores, medusae and siphonophores. Marine Biology 133:571–582.

Harbison, G. R. 1984. On the classification and evolution of the Ctenophora. Pages 78–100 *in* S. C. Morris, J. D. George, R. Gibson, and H. M. Platt, editors. The origins and relationships of lower invertebrates. Oxford University Press, Oxford, UK.

Harbison, G. R. 1996. Ctenophora. Pages 101–147 *in* R. Gasca and E. Suárez, editors. Introducción al Estudio del Zooplancton Marino. ECOSUR, Chetumal, Mexico.

Harbison, G. R., and L. P. Madin. 1982. Ctenophora. Pages 707–715 *in* S. P. Parker, editor. Synopsis and classification of living organisms. McGraw-Hill, New York.

MacGinitie, G. E. 1955. Distribution and ecology of the marine invertebrates of Point Barrow, Alaska. Smithsonian Miscellaneous Collections 128(9): 1–201.

Matsumoto, G. I., and B. H. Robison. 1992. Kiyohimea usagi, a new species of lobate ctenophore from the Monterey Submarine Canyon. Bulletin of Marine Science 51:19–29.

Mills, C. E. 1987. Phylum Ctenophora. Pages 79–81 *in* E. N. Kozloff, editor. Marine invertebrates of the Pacific Northwest. University of Washington Press, Seattle.

Mills, C. E. 1996. Additions and corrections to the keys to Hydromedusae, hydroid polyps, Siphonophora, stauromedusan Scyphozoa, Actiniaria, and Ctenophora. Pages 487–491 *in* E. N. Kozloff, editor. Marine invertebrates of the Pacific Northwest, with revisions and corrections. University of Washington Press, Seattle.

Mills, C. E. 1998–2000. Phylum Ctenophora: list of all valid species names. Electronic internet document available at http://faculty.washington.edu/cemills/Ctenophores.html. Published by the author, web page established March 1998, last viewed 30 June 2001.

Nybakken, J. W. 1993. Marine biology: an ecological approach, 3rd edition. Harper Collins College Publishers, New York.

Nybakken, J. W. 1997. Marine biology: an ecological approach, 4th edition. Addison-Wesley Educational Publishers, New York.

Wrobel, D., and C. E. Mills 1998. Pacific coast pelagic invertebrates: a guide to the common gelatinous animals. Sea Challengers and the Monterey Bay Aquarium, Monterey, California.

PART III
Index

Bold numbers indicate illustrations

B

C

D

G

H

I

M

N

Q

T

X

Z

Portfolio of Cnidarian Diversity

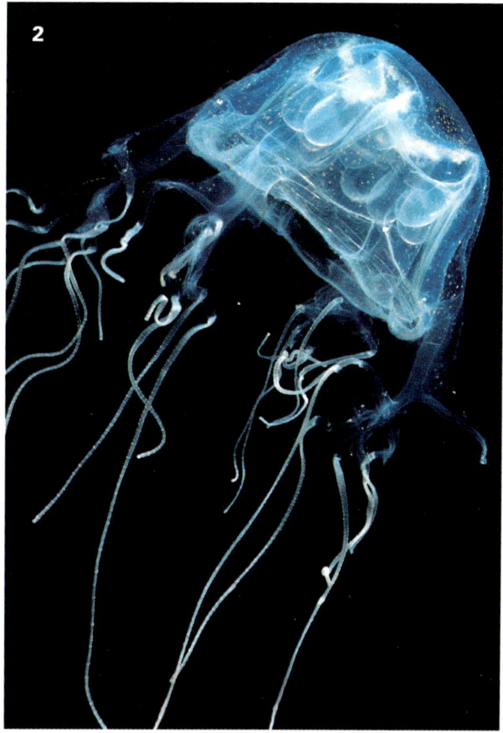

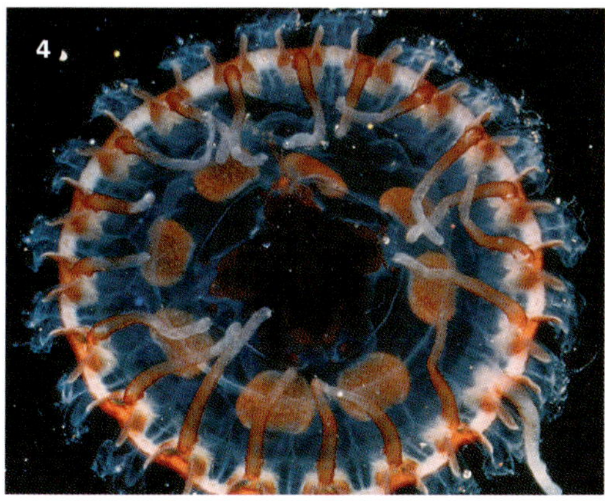

1 A carybdeiid cubozoan *Tamoya haplonema*, Class Cubozoa (A. Migotto)

2 A chirodropid cubozoan *Chiropsalmus quadrumanus*, Class Cubozoa (A. Migotto)

3 A stalked jellyfish *Haliclystus stejnegeri*, Order Stauromedusae (R. J. Larson)

4 A crown jellyfish *Atolla vanhoeffeni*, Order Coronatae (S. H. D. Haddock)

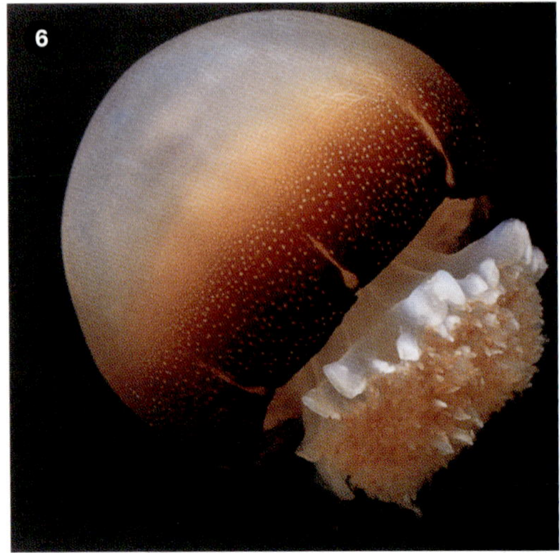

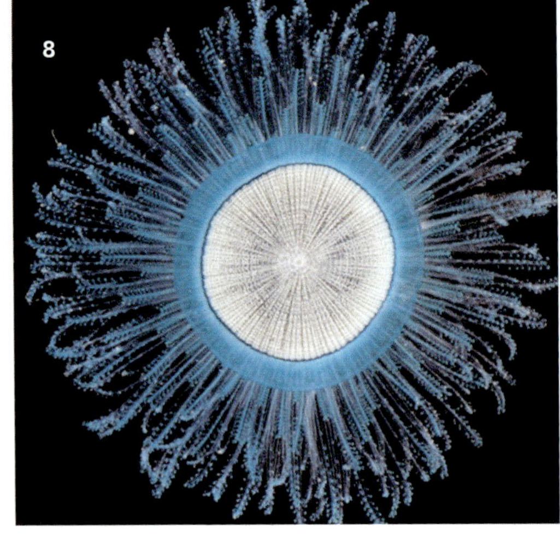

5 Sea nettle *Chrysaora quinquecirrha*, Order Semaeostomeae (R. J. Larson)

6 Cannonball jellyfish *Stomolophus meleagris*, Order Rhizostomeae (R. J. Larson)

7 Club hydroid *Clava multicornis*, Order Anthoathecatae (R. J. Larson)

8 Blue button *Porpita porpita*, Order Anthoathecatae (C. E. Cutress)

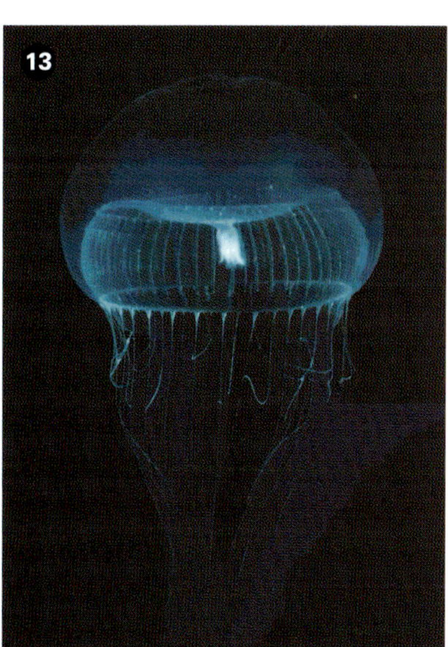

9 By-the-wind sailor *Velella velella*, Order Anthoathecatae (R. J. Larson)
10 *Cladonema radiatum*, Order Anthoathecatae (R. J. Larson)
11 *Aglauropsis aeora*, Order Limnomedusae (D. Wrobel)
12 Stinging hydroid *Macrorhynchia philippina*, Order Leptothecatae (D. R. Calder)
13 Water jellyfish *Aequorea victoria*, Order Leptothecatae (S. H. D. Haddock and MBARI)

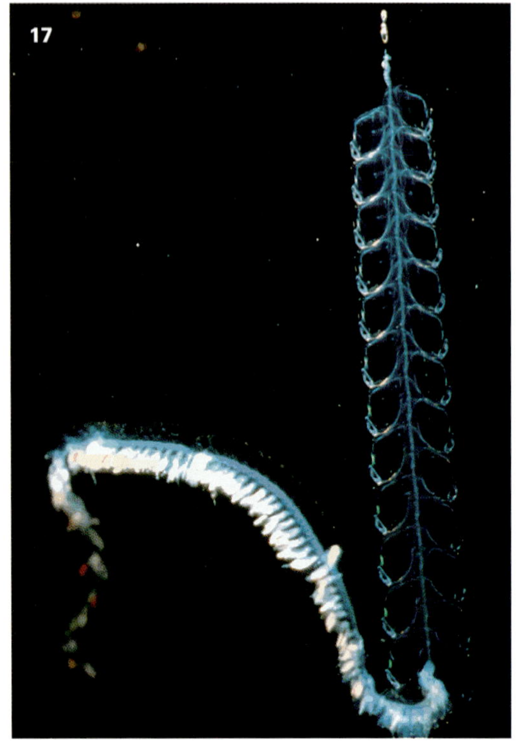

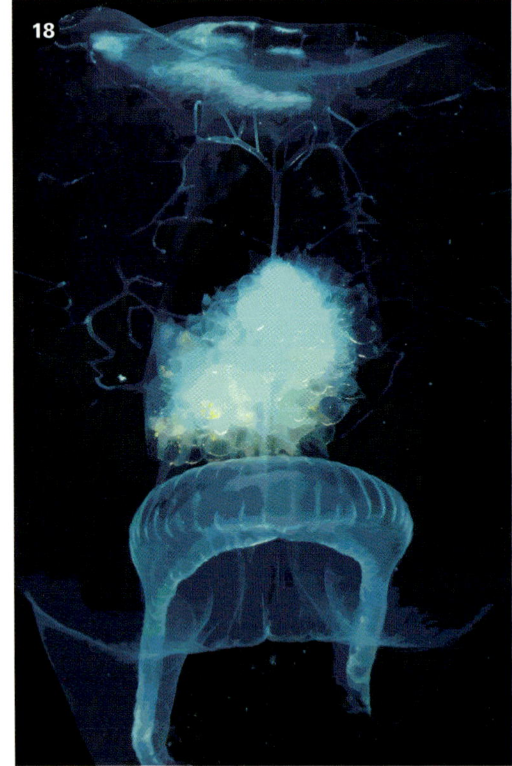

14 *Pegantha* sp., Order Narcomedusae (D. Wrobel)

15 *Benthocodon pedunculata*, Order Trachymedsae (C. Mills)

16 Portuguese man-of-war *Physalia physalis*, Order Cystonectae (siphonophore)

17 *Bargmannia* sp., Order Physonectae (siphonophore) (S. H. D. Haddock)

18 *Praya dubia*, Order Calycophorae (sipnophore) (S. H. D. Haddock and MBARI)

19 Black wire coral *Stichopathes lutkeni*, Order Antipatharia (C. E. Cutress)
20 A tube-dwelling anemone *Cerianthus* sp., Order Ceriantharia (G. K. Jensen)
21 Deep-water sea fan *Iciligorgia schrammi*, Order Alcyonacea (C. E. Cutress)
22 A sea pen *Pennatula aculeata*, Order Pennatulacea (R. J. Larson)
23 *Bunodosoma granuliferum*, Order Actiniaria (C. E. Cutress)

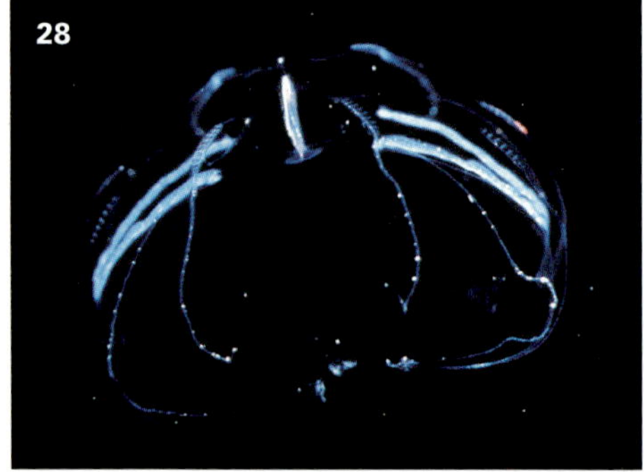

24 *Zoanthus* sp., Order Zoanthidea (K. Larson)
25 Ctenophore *Hormiphora californensis*, Order Cydippida (D. Wrobel)
26 *Discosoma sanctithomae*, Order Corallimorpharia (C. E. Cutress)
27 Elkhorn coral *Acropora palmata*, Order Scleractinia (W. C. Jaap)
28 Ctenophore *Thalassocalyce inconstans*, Order Thalassocalycida (D. Wrobel)

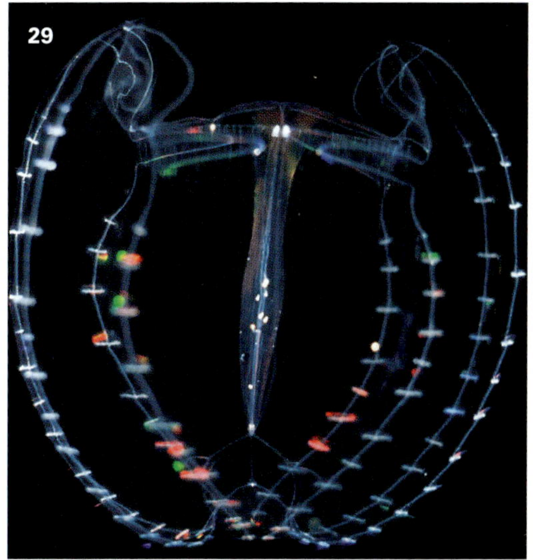

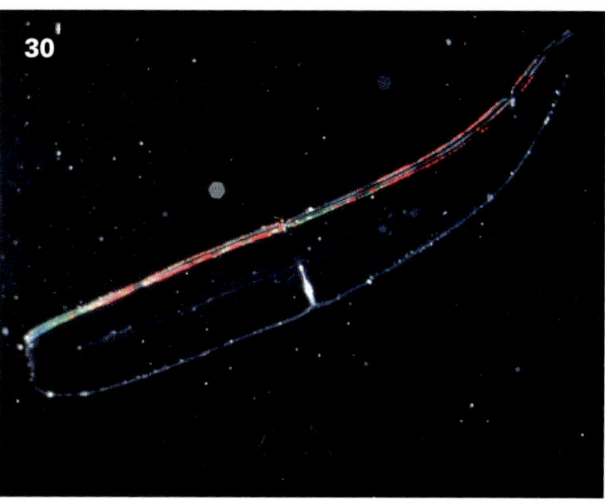

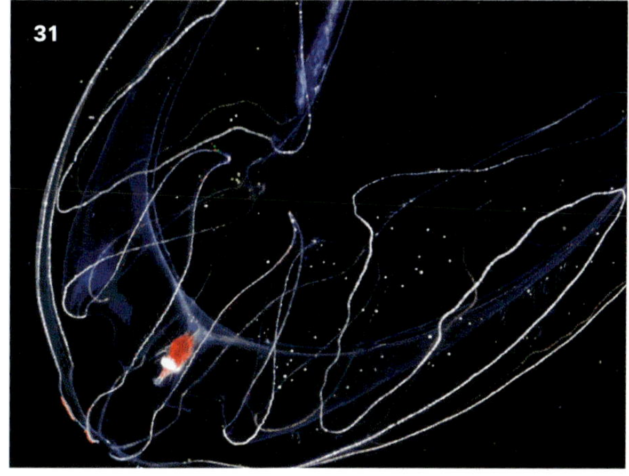

29 Ctenophore *Deiopea* sp., Order Lobata (S. H. D. Haddock)
30 Ctenophore *Velamen parallelum*, Order Cestida (D. Wrobel)
31 Ctenophore *Bathycyroe* sp., Order Lobata (S. H. D. Haddock)
32 Ctenophore *Beroe forskalii*, Order Beroida (S. H. D. Haddock)